Giovanni Paolo Lomazzo's *Trattato dell'Arte della Pittura*

Nuncius Series

Studies and Sources in the Material and Visual History of Science

VOLUME 1

The titles published in this series are listed at *brill.com/nuns*

Giovanni Paolo Lomazzo's *Trattato dell'Arte della Pittura*

Color, Perspective and Anatomy

By

Barbara Tramelli

BRILL

LEIDEN | BOSTON

Cover illustration: Giovanni Paolo Lomazzo, *Self portrait as a young boy*, c. 1570, Vienna, Kunsthistorisches Museum.

Library of Congress Cataloging-in-Publication Data

Names: Tramelli, Barbara, author.
Title: Giovanni Paolo Lomazzo's Trattato dell'arte della pittura : color, perspective and anatomy / by Barbara Tramelli.
Description: Boston : Brill, 2017. | Series: Nuncius Series : studies and sources in the material and visual history of science ; volume 1 | Includes bibliographical references and index.
Identifiers: LCCN 2016036664 (print) | LCCN 2016036857 (ebook) | ISBN 9789004330252 (hardback : alk. paper) | ISBN 9789004330269 (E-book)
Subjects: LCSH: Lomazzo, Giovanni Paolo, 1538-1600--Knowledge and learning. | Lomazzo, Giovanni Paolo, 1538-1600. Trattato dell'arte della pittura. | Communication in science--Italy--Milan--History--16th century. | Communication in art--Italy--Milan--History--16th century. | Milan (Italy)--Intellectual life--16th century.
Classification: LCC ND623.L7284 T73 2017 (print) | LCC ND623.L7284 (ebook) | DDC 759.5--dc23
LC record available at https://lccn.loc.gov/2016036664

Typeface for the Latin, Greek, and Cyrillic scripts: "Brill". See and download: brill.com/brill-typeface.

ISSN 2405-5077
ISBN 978-90-04-33025-2 (hardback)
ISBN 978-90-04-33026-9 (e-book)

This book is printed on acid-free paper and produced in a sustainable manner.

Contents

Acknowledgements

First of all I wish to thank Professor Marco Beretta and Professor Sven Dupré for agreeing to publish this manuscript in the 'Nuncius' series.

I also wish to thank the Max Planck Institute for the History of Science in Berlin for supporting my research and in particular the Max Planck research group "Art and Knoweldge in Pre-Modern Europe" directed by Professor Sven Dupré for having me as pre-doctoral fellow in the past years. Colleagues there with whom I shared my research have been helpful in so many ways, for which I am enormously grateful. I wish to thank my co-supervisor, Professor Erna Fiorentini, for her kindness and help. I am grateful to all the scholars whom I had the privilege to meet and who generously shared their expertise with me during my PhD studies: Professor Karin Leonhard, Professor Stephanie Leitch, Professor Lorraine Daston, Professor Claudia Swan, Dr Sylvie Neven, Dr Marlise Rijks, Dr Marjolijn Bol, Professor Pietro Roccasecca, Dr Allison Ksiazkiewicz, Dr Sean Nelson, Dr Michael Bycroft, Professor Erma Hermens, Professor Karin Gludovatz, Professor Wolf-Dietrich Löhr, Dr Ulrike Boskamp, Dr Sheila Barker, Dr Ulrike Kern, Dr Sietske Fransen, Professor Alexander Marr, Dr Tawrin Baker, Professor Giulio Bora, Professor Claire Farago, Dr Mira Becker, Professor Rossana Sacchi, Professor Silvio Leydi, Professor Richard S. Field, Dr Thijs Weststeijn, Dr Nadia Baadj, Dr Elaine Leong and Dr Skuli Sigurdsson. Thank you all so much. Finally, for their support and friendship, a special thank goes to the MPI research group staff members Gina Grzimek, Karin Weninger, Giulia Simonini, Celine Camps and Linda Oldenburg.

Outside of Berlin, I would like to express my gratitude to the amazing scholars at the Warburg Institute, and especially to Professor Jill Kraye, Professor Elizabeth McGrath, Professor Charles Hope and Dr Guido Giglioni, who were always ready to provide help and guidance.

Warm thanks are due to Alexandra Challenger, Professor Jill Kraye and Professor Stephanie Leitch for checking my English. I am grateful to Dr. Mauro Pavesi for kindly sharing his work and his picture of the Foppa chapel.

I also want to thank so many friends, whose presence, love and constant support were vital in order to keep my spirits high, especially during long German winters: in Piacenza, my oldest friends and practically family in Piacenza Camilla, Simona, Serena; in London, Laura and Micol; in Milan, Sara, Elena, Virginia, Erika, Jacqueline; and in Berlin, Maria, Rick, Sean, Jack and many others, who have made all the difference to my life there.

Needless to say, my deepest gratitude goes to my family, who have always supported me and trusted in my decisions. A special thanks to my parents,

Anna Maria and Giorgio, who followed me every step of the way with unconditional love and care, and to my brother Filippo, his wife Alessandra and their wonderful family. I dedicate this book to my beloved nephews, Giacomo and Matteo.

Editorial Principles

I have quoted Lomazzo's works, apart from the *Rime* and the *Rabisch*, from the critical edition by Roberto Paolo Ciardi, *Scritti sulle Arti*, Pisa 1973. References to the 1584 edition of Lomazzo's *Trattato* are added in the footnotes in brackets. Unless otherwise indicated, quotations from treatises on art by other writers are taken from the first editions. Obvious misprints from Italian texts have been silently corrected. All translations are my own, unless otherwise indicated, and I have provided the original language in the footnotes. Although I consulted at Richard Haydocke's English translation of Lomazzo's treatise (1598), I did not use his version, since it was not sufficiently accurate for my purposes.

Full details of the works cited are given in the bibliography. In footnotes, they are indicated with the author's name, title, place and date of publication.

List of Illustrations

Introduction: Aims, Sources and Methodology*

The aim of this monograph is to investigate the education of artists and, in particular, to explore the pictorial knowledge of the sixteenth-century Milanese painter and writer on art Giovanni Paolo Lomazzo (1538–1592). I examine his writings (with special attention to the *Trattato dell'Arte della Pittura*, published in 1584), with an eye to understanding how the context in which he was involved shaped the notions conveyed in the *Trattato*, and how the pictorial knowledge in the artist's workshop was classified and transformed into a treatise. Lomazzo's books testify to the redefinition of the status of artists by means of compiling and publishing writings on the science of painting. His literary works followed the example of previous and contemporary writers on art who, mainly – but not exclusively – used the treatise format to categorize and systematize the different parts of painting. The interest of artists in writing treatises has usually been regarded as closely bound up with the intellectualization of the painter's task, that is, the notion that the appropriation of knowledge through books became more important for artists of this period in defining their pictorial subjects (*invenzioni*) and in forming an appropriate artistic vocabulary. As we shall see, however, Lomazzo's writings testify to his desire to communicate to readers not only theoretical ideas but also information on the practical aspects of the painter's endeavor, including suggestions about color, perspective and anatomy, and employing a terminology which reflected the reality of the artist's workshop.

Lomazzo's life and literary works provide the basis for this detailed study of him as an artist, writer and academician. He was responsible for the first "methodical manual" on the art of painting (Ackerman 1967), and the range of his interests in different subjects, both related and unrelated to his artistic activity, indicates an attempt to extend his reputation beyond that of a painter.

My main aim is to set his literary works in context, by taking into account the environment in which they were created as an essential background for the material that he assembles in his books. This study expands on the excellent work done forty years ago by Roberto Paolo Ciardi in identifying Lomazzo's sources and examining his treatment of various elements of painting in con-

* This study is a revised version of my doctoral thesis entitled "The Art of Writing, the Writing of Art: Giovanni Paolo Lomazzo's Context, Connections and Influences in his *Trattato dell'Arte della Pittura*", defended at the department of Art History at the Free University of Berlin (29/09/2015). I wish to thank my supervisor, Professor Sven Dupré, for each insightful discussion I had with him during my PhD years.

 | DOI 10.1163/9789004330269_002

nection with the state of the arts at the time.[1] In doing so, I investigate what kind of information and *exempla* Milanese artists studied and discussed among themselves, and I show how the so-called 'tacit knowledge' acquired in the artist's workshop was codified and shared at the time. My aim is to present Lomazzo's literary achievements as the result of his studies, interests and connections both with artists and with men of letters, and to evaluate the extent to which his knowledge was the product of reading and/or the consequence of interactions and conversations with the Milanese literary and artistic communities.

There have been many studies on the education of artists in Renaissance Italy, from Rudolf Wittkower's *Born Under Saturn* to the work of scholars such as Charles Dempsey, Jan Bialostocky, Francis Ames-Lewis, and Elizabeth McGrath, who have attempted to determine the educational level of artists in the early modern period and to discover which books were possessed, read, and used by painters, sculptors and architects.[2] Recently, *The Artist as Reader*, a collective volume edited by Heiko Damm, Michael Thimann and Claus Zittel, expanded on previous about research artists' libraries and positioned the activity of "reading as the greatest proof of refinement when viewed within the context of the social climb of the visual artist," taking into account a variety of case studies and different aspects of how artists' education, language skills, and use of libraries were related to the issue of their self-fashioning.[3] Close attention has always been paid to inventories, and the findings in this field have proven to be to some extent unexpected.[4]

Lomazzo's writings –the *Trattato dell'Arte* (1584) and the *Idea del Tempio della Pittura* (1590), along with the short *libello* dedicated to Duke Ferdinando de' Medici entitled *Della Forma delle Muse* (1591) – are his most prominent

1 Roberto Paolo Ciardi, *Scritti sulle Arti*, 2 vols., (Pisa: Marchi & Bertolli, 1973).

2 Rudolph and Margot Wittkower, *Born under Saturn* (London: Weidenfeld & Nicolson, 1963); Charles Dempsey, "Some Observations on the Education of Artists in Florence and Bologna during the Later Sixteenth Century," *The Art Bullettin*, December 1980, *LXII*:552–569; Jan Bialostocki, "Doctus Artifex and the Library of the Artist in XVIth and XVIIth Century," in *De Arte et Libris, Festschrift Erasmus 1934–1984*, edited by Abraham Horodisch (Amsterdam: Erasmus Antiquariaat en Boekhandel, 1984), pp. 11–22; Francis Ames-Lewis, *The Intellectual Life of the Early Renaissance Artist* (London and New Haven: Yale University Press, 2002); Elizabeth McGrath, "Artists and Mythographic Handbooks: Some Evidence of Use and Ownership," *Warburg Institute Colloquia*, 2009, *XIV*:389–419.

3 Heiko Damm, Michael Thimann, and Claus Zittel (eds.), *The Artist as Reader. On Education and Non-Education of Early Modern Artists* (Boston and Leiden: Brill, 2010).

4 See especially Lex Hermans, "Reading Rhetoric: Oratory in Gian Paolo Lomazzo's Treatises on the Art of Painting," in *The Artist as a Reader*, pp. 241–258.

achievements in the genre of "art theory." Other writings such as his *Libro de' Sogni* (1563), *Rime* (1587), and *Rabisch* (1589) do not have painting as their main subject, although they should still be taken into consideration in order to understand Lomazzo's literary background and especially his opinion on different topics beyond the realm of art. The *Trattato* and *Idea* have often been quoted and used by art historians to support different claims. Since their "rediscovery" by Erwin Panofsky (1924), they have been regarded as one of the most complete attempts to define the structure of painting from a Neoplatonic approach and in line with the new concept of 'Idea' (Tea 1939, Blunt 1940, Klein 1959, 1961, Ackerman 1967, Kemp 1987, Bora 1989, Deswarte-Rosa 1991, Manegold 2004). These treatises have been seen as testimonies to the rising status of painting during the sixteenth and early seventeenth centuries (Ciardi 1965, 1973, Isella 1993, Barasch, 1960, 1978, 1997, Williams 1997); and they have also been used as evidence for the reception of Leonardo's works (Bora 1971, 1980, 1987, 1998, Kwakkelstein 1993, Meijer 1998, Porzio 1998, Berra 1993, 1998, 2009).

While these books have been regarded, up to the present day, as a milestone in early Mannerist art theory, they have rarely been analyzed in terms of the context in which they were written. The benefit of such an approach is that it broadens the scope of the investigation to include men of letters and of science who were in contact with Lomazzo and with his circle of artists. A praiseworthy move in this direction was the exhibition *Rabisch: Il Grottesco nell'Arte del Cinquecento, L'Accademia della Val di Blenio, Lomazzo e l'Ambiente Milanese*, held at the Museo Cantonale d'Arte in Lugano in 1998. Scholars such as Giulio Bora, Francesco Porzio and Dante Isella outlined the environment to which Lomazzo belonged, with emphasis on the culture of the Counter-Reformation promoted by Carlo Borromeo, on whom there is a vast literature. More recently, the exhibition *Arcimboldo: Milanese Painter between Leonardo and Caravaggio*, held in 2011 in the Palazzo Reale of Milan, not only presented the context and cultural life of the city but also hinted at the economic connections at the time between Milan and the rest of Europe. Nonetheless, further information is needed about the art scene in Lombardy during the second half of the sixteenth century. Moreover, even though Lomazzo has been extensively studied as a fundamental source for Italian Renaissance art history, he has not previously been approached from outside the perspective of art historical research on the period. As Roberto Ciardi and Gerald Ackerman have pointed out, the main issues in dealing with Lomazzo's writings are his disorganized and in many parts obscure style, and the poor condition of the texts themselves, which are at various points difficult to interpret.[5] In addition, Lomazzo,

5 Gerald Ackerman, "Lomazzo's Treatise on Painting," *The Art Bullettin*, December 1967, XLIX. 4:317–326.

as a source of information – like Giorgio Vasari – has sometimes been regarded as unreliable: too intent on praising the works and skills of Northern painters to be an objective judge of contemporary Italian artists.[6]

Despite these difficulties, a deeper enquiry into the nature, the circumstances and the sources of Lomazzo's literary works is necessary to gain a more comprehensive understanding of, firstly, the types of skill and knowledge, as well as the discussions held by artists belonging to Lomazzo's circle – among themselves and with men of letters – and, secondly, how different bodies of knowledge were used by him to systematize painting.

A crucial issue for the interpretation of Lomazzo's writings is that the readership for which they were intended has yet to be identified. It is fairly reasonable to assume that the *Rabisch* was supposed to be read in the context of the *Accademia de la Val di Blenio* and that it was most probably not aimed at a wider audience (not many people, outside of Milan or the academy itself, would have understood the fictitious *Bregnese* dialect in which most of the book is written). The intended readership of the unpublished *Libro de' Sogni*, however, remains unclear. As far as the *Trattato* (along with the *Idea*) is concerned, we know that it was first published in 1584 by Paolo Gottardo Pontio and went through three editions in the same year, followed by another in 1585. Had the book not been of interest to a certain type of reader in Milan, it is unlikely that four editions would have been produced within such a short space of time; this is consistent with the argument that the main aim of these re-editions was to correct the faulty editing in order to make the book more appealing to the widest possible readership.[7]

In the 'General Conclusions' of this study I refer to a few of the people (both artists and men of letters) who we know, mainly through inventories and exchanges of letters, possessed Lomazzo's books. I have not presented, however, a detailed account of the book's *fortuna*, which I had initially hoped to do; this is because information about who might have owned and read the book proved to be too scattered to determine whether or not it was used by artists at the time. Nevertheless, the characterization of the *Trattato* as an "editorial

---

6 See, e.g., Martin Kemp, *The Science of Art: Optical Themes in Western Art from Brunelleschi to Seurat* (New Haven: Yale University Press, 1990), p. 73, on Lomazzo's "chauvinistic attempt" to praise Vincenzo Foppa's skills in perspective.

7 In the first edition, the title was *Trattato dell'Arte della Pittura*, while in the second edition of the same year it became *Trattato dell'Arte della Pittura, Scoltura et Architettura*. A chapter was abandoned in the second edition and then reintroduced in the third edition, while some errors in the first edition were corrected in the 1585 edition. Some aspects of the editorial process are still unclear; see Gerald Ackerman, *The Structure of Lomazzo's Treatise on Painting* (Princeton: Princeton University Press, 1964).

failure" should be, if not reconsidered, at least toned down. Of greater interest is for this study the readership for which Lomazzo intended the treatise. He states in various passages of his writings that his main desire is to assist artisans and painters who are starting to learn the fundamentals of painting, though this claim should be taken with a grain of salt.[8] Ultimately, we can divide his readership into two groups: those for whom he explicitly wrote the book, that is, artists, and those who were probable readers and buyers, that is, "art lovers", or the *professori del disegno*.

After declaring that he wants his literary works to be useful for artists in their profession, as many other writers on art had done before him, Lomazzo states that, as a painter himself, he is aware of two distinct difficulties faced by painters when dealing with his books: firstly, they have limited knowledge of philosophical precepts; and, secondly, it is impractical for them to read and consult books in their workshops. To overcome both difficulties, he adopted what seems to be an even more unpractical solution: the addition of a sixth and seventh book. Thus, in the *proemio* to the *Trattato*, he writes:

> Considering that not all who start to learn this art will be able to pick the fruit of this tree, as it is too high (I mean that in the first five books we treat the essential art and rules of painting in general, which are not familiar to our senses, so it is not easy for everyone to discern which general principle applies to this or that specific thing), I, desiring profit and usefulness for those who start to learn this art, will add a sixth book, where I will treat practically what I explained theoretically in the first five books; and this is also the order of doctrine, that after theory comes practice.[9]

8 Giovanni Paolo Lomazzo, *Trattato dell'Arte della Pittura,* (Milan: Paolo Gottardo Pontio, 1584), p. 14. Anthony Blunt, *Artistic Theory in Italy, 1450–1600* (Oxford: The Claredon Press, 1940), pp. 137–159; Robert Williams, *Art, Theory and Culture in Sixteenth-Century Italy* (Cambridge: Cambridge University Press, 1997), pp. 123–187.

9 Ciardi, *Scritti*, II, p. 23: "Ma considerando che non tutti quelli che cominceranno ad apprendere quest'arte sapranno cogliere il frutto di quest'arbore, per essere troppo alto (voglio dire che trattandosi in questi primi cinque libri de le arti essenziali e principali de la pittura generalmente et essendo le cose generali troppo discoste dal nostro senso, onde tutti non sapranno discernere a quale universale questa cosa particolare o quest'altra si sottoponga), io che sommamente desidero il profitto e l'utilità eziando di quelli che cominciano a imparare quest'arte, ho voluto aggiungere un sesto libro nel quale tratterò pratticamente quello che ne i cinque libi si insegna teoricamente: essendo anco questo l'ordine della dottrina, che doppo la teorica seguiti la pratica." (Lomazzo, *Trattato*, p. 15, *Divisione di Tutta l'Opera*).

Lomazzo then states that he specifically added a seventh book, in which the painter could find the *storie* which were necessary to his work:

> Because the young practitioner needs not only the rules of art, but also the precepts of *giudicio* and *prudenza*, immediately before the discourse on the practice, I have prefixed a compendium of the rules of art, together with a collection of the precepts of discretion and judgement which an artisan ought to use in painting.[10]

He did so by "reading and re-reading an infinite number of books and not having any thought for my own benefit and well-being, in order to be of service and use to the people exercising my profession."[11] He expresses a similar aim in his dedication of the *Idea*: "having in mind the benefit that could follow, I have demonstrated to others the rapid and plain way to imitate nature, in which the art of painting consists. This is something that few can understand without the illumination of rules."[12] Lomazzo says that he wants painters to use his book as a sort of compendium in which they can find everything on the art of painting, a book (he states more than once) designed for practical consultation. It has, indeed, been argued that the *Trattato* was a practical "handbook for the artist, the critic, and the amateur who might wish to avail himself of the accumulated store of Italian experience in the field."[13] Yet, although the *Trattato* was divided into seven books for easier reading, it amounted to a total of seven hundred pages and, therefore, would have been quite unwieldly to consult in an artist's workshop.

---

10 Ciardi, *Scritti*, II, p. 23: "E perchè quelli che praticamente cominciano ad adoperare, non solo hanno bisogno de le regole dell'arte, ma ancora de i precetti del giudicio e de la prudenza, in questo istesso libro, avanti ch'entrassi a trattare de la pratica, ho premesso un compendio di regole de l'arte, insieme con una raccolta di precetti de la prudenza e giudicio che ha di avere l'artefice nel dipingere. Perchè non basta al pittore che dipinga bene, ma gli si ricerca anco che dipinga con prudenza e giudicio." (Lomazzo, *Trattato*, pp. 15–16).

11 Ciardi, *Scritti*, II, p. 24: "A che fare è stato necessario leggere e rileggere infiniti libri e non avere alcun riguardo a l'utile e comodo mio privato per apportar utile e servizio a gl'uomini de la mia professione." (Lomazzo, *Trattato*, p. 16).

12 Ciardi, *Scritti*, I, p. 244: "Et in oltre pensando al giovamento che potea seguire, co'l dimostrare altrui la via spedita e piana d'imitare e come emular la natura, in che consiste tutta l'arte de la pittura. Cosa che da pochi senza il lume delle regole e dei precetti può essere intesa." (Lomazzo, *Idea* (Milan: Paolo Gottardo Pontio, 1590), Dedication to Don Filippo d'Austria).

13 Frederick Hard, "Some Interrelations Between the Literary and the Plastic Arts in 16th and 17th Century England," *College Art Journal*, 1951, *X.3*: 233–243, p. 235.

Lomazzo's stated aim of producing a treatise which was "useful for painters" and which they should read could be seen as part of his attempt to raise the social stature of artists, and this was connected to the goal of placing painting among the liberal arts. As we have seen, although practicing artists were his explicitly intended readership, "art lovers" or *professori del disegno* were probable readers and purchasers of the book – though in fact, there is evidence that some artists also possessed the book.[14]

Lomazzo's literary achievements reflect the tendency of sixteenth-century artists to consider practical and theoretical knowledge as both complementary and necessary for painters. This is clearly indicated at the beginning of the eighth chapter of *Idea* (*Delle scienze necessarie al pittor*e- "On the sciences necessary for the painter"),[15] where he draws a distinction between the two approaches to painting, practical and theoretical, stating that both were essential for a painter who wants to reach perfection.[16] He argues that to achieve this an artist must not only study literature and poetry, but also geometry, mathematics, astrology, and perspective.[17] In this sense, Lomazzo does not distance himself from earlier writers on art (starting with Leon Battista Alberti) who maintained that a "complete" artist should master both the practice and the theory of painting. What is substantially different is how he defines practice, since his notion of *pratica* is not immediately clear, and, as Moshe Barasch noted, "Practice (*in Lomazzo*), it turns out, is not less theoretical than theory."[18] Although I do not completely agree with this view, Lomazzo does embed

14 See 'General Conclusions'.

15 Ciardi, *Scritti*, I, p. 271.

16 Ibidem: "Due sono le vie di operare nella pittura, una di pratica, l'altra di teorica. Per pratica opera colui che senza sapere il fondamento e la ragione di quello che fa, ha solamente una certa facoltà, ch'egli ha acquistato con un lungo esercitarsi, o si regge solamente dietro ad alcun esempio. Ma per teorica opera quello che sa mostrar con ragione di proporzionati effetti, le perdite e i ravvolgimenti dei corpi, e tutto quello che si può far col pennello, e appresso gli sa esplicare con parole, e insegnargli con ordine, con chiarezza, e con facilità ad altri." (Lomazzo, *Idea*, p. 28).

17 Ciardi, *Scritti*, I, p. 273: "Con la geometria verrà a conoscere i corpi perfetti e regolati, con le loro proporzioni e misure, che sono i fondamenti delle trasferizioni, in che pende tutta l'arte; con la prospettiva, che è il cuore della geometria, le ombre, i lumi, i raggi, gli scorti e finalmente tutte quelle parti che, ingannandogli occhi nostri, ci fanno vedere quello che non è. Con l'aritmetica le proporzioni, le armonie e le convenienze dei corpi, per numeri e quantità, perciochè co'l numerar le parti minime con le maggiori, si vengono a formar le pitture giuste e belle e non fatte a caso come sono quelle di coloro che sono privi di questa cognizione tanto necessaria." (Lomazzo, *Idea*, p. 29).

18 Moshe Barasch, *Theories of Art* (New York and London: New York University Press, 1985), p. 274.

theoretical notions in what he declares to be the most "practical" book of the treatise, that is, *Della pratica della pittura*. One example, as we shall see, is his complex system of colors in connection with planets, elements, and temperaments, which, although there are some hints about it in book III (*Del colore*), is thoroughly explained only in the sixth book of the treatise, devoted to the practice of painting. On the other hand, recipes and technical suggestions on how to paint on different surfaces are placed in what he declares to be his "theoretical book" on color, *Del colore*. Similarly, his practical suggestions on how to use perspective are listed in the book *Della prospettiva*.

These examples show that his apparently strict categorization of the parts of painting is, in some respects, misleading, since, despite the author's statements, there is no clear-cut division between the topics presented in the book. For instance, we find interesting material on color and perspective in the book of light, and vice versa, or anatomical information in the book of perspective, and so on. The discourse becomes even more complex if we also take into account the later *Idea del Tempio della Pittura*, the relationship of which to the *Trattato* has been thoroughly discussed by scholars from Robert Klein onwards.[19] These inconsistencies in the organization of topics have usually been explained by referring to the undefined eye disease which struck Lomazzo at the age of thirty-three (some have speculated that it was a form of early onset cataracts) and which prevented him from continuing his work as a painter.[20] In my view, even though this handicap made it impossible for him to continue painting, he was still able to read the "four thousand and more papers" that he claimed, in the autobiography at the end of the book *Rime*, to have collected.[21] Lomazzo states that after he lost the "dear light" of his eyes, he turned to writing; however, it is generally agreed that he always had an interest in writing, which he started pursuing from an early age, as he writes in the autobiography and as the manuscript of the *Libro de' Sogni* (1563) testifies.[22] His possession of a large number of papers, along with his collection of writings by other artists, including most notably Leonardo, shows that he had a keen interest in the theory of painting long before his supposed blindness compelled him to give up his

19 Robert Klein, *Idea del Tempio della Pittura*, 2 vols. (Florence: Istituto Nazionale di Studi sul Rinascimento, 1974).

20 Lomazzo talks about this problem in different places, especially in his autobiography at the end of the book *Rime* (Milan: Paolo Gottardo Pontio, 1587), p. 538: "E perdei l'amata e cara luce/Che mi fece restar fuor di me stesso."

21 Giovanni Paolo Lomazzo, *Rime*, p. 542: "Di mano di pittore vari disegni /Raccolsi, et carte rare e principali,/Si de l'Italia quanto forastiere; /Che a quattro mille giungean tutte scelte."

22 British Library, London, Add MS 12196, Giovanni Paolo Lomazzo, *Libro de' Sogni* (1563).

trade.[23] Nevertheless, the process of writing of the book was certainly affected by his failing sight, and he must have secured the services of some "silent helpers" to help him compile the impressive amount of material contained in his books. To judge from his own words, Lomazzo's literary activity was intended to be a complementary task to his activity as an artist; for he states in the *proemio* to the *Idea* that he would have liked to have added an illustrative compendium of drawings to his books, which "would have been very useful not only to practical men (i.e., artists), but also to theorists."[24] It is these *studiosi* whom Lomazzo addresses at the end of the book:

> The *studiosi*, praising more the substance of things than the sweet sound that pleases the ear, should take some time to read this treatise, as it will be useful and worthy, and they will apply the precepts that they have learnt here, and they will praise me, if not for *ingegno*, at least for my diligence and for my honest desire to be of benefit to the world and for the effort that I made in achieving this.[25]

Whether the word *studiosi* here refers to men of letters or to the artists whom Lomazzo hoped would buy and read the treatise is unclear, though, in my view, he more likely had in mind men of letters beyond his own artistic circle.

As has been pointed out in the past, Lomazzo's purpose – or one of his purposes – in publishing the treatise was to re-evaluate art from Northern Italy, and especially from Milan, in order to establish an alternative canon to the Florentine literature on art embedded in Vasari's *Vite*. It is for this reason that, in the conclusion to his book, Lomazzo declares that he has discussed the parts of painting in a "familiar and ordinary way, without ornament, also including

---

23 Lomazzo, *Rime*, p. 531.

24 Ciardi, *Scritti*, I, p. 244: "Era mio proponimento ancora, e di già gli aveva dato principio, di disegnar le figure per le quali si potesse più chiaramente comprendere tutta la ragione di operare e mettere in prattica quanto per via di regole e precetti aveva insegnato. Il che sarebbe stato di grandissima utilità, non solo a puri prattici, ma anco a i teorici."

25 Ciardi, *Scritti*, II, p. 589: "Gli studiosi, adunque, pregiando più la sodezza delle cose che un dolce suono che gli lusinghi l'orecchio, non restino d'impiegare alle volte qualche ora che gli avanzi in leggere questo Trattato che senza dubbio ne riporteranno utile et onore osservando quei precetti che quivi averanno apparato, e loderanno me, se non dell'ingegno, almeno della diligenza e di questo onestissimo desiderio che ho avuto di giovare al mondo e dello sforzo che ho fatto per conseguirlo." (Lomazzo, *Trattato*, p. 680).

many words that are not common (*meno approvate*) and that Tuscan people do not use."[26] This statement is consistent with his choice of a terminology belonging to the workshop, as he continues: "Concerning the words that are not common, these belong so closely to this art and consequently are so meaningful among painters, that they could not in any way be left out, since my aim is to be understood."[27] This *captatio benevolentiae* is intended to justify Lomazzo's use of the simple language of painters, as well as his inclusion of specific terms familiar to and comprehensible by them. Ironically, his treatise was later considered by scholars to be one of the most complex and difficult works on art to understand, because of the often inconsistent structure of his discourse – consistency was not a priority for him – and the ambiguous and far from fluent style of his vernacular *lessico*.

According to a recent translation of the *Idea* by Jean Julia Chai: "Lomazzo usually accumulates run-on sentences with subordinate clauses, tense changes, ambiguous pronouns and erratic word order", which makes it even harder to understand his line of thoughts.[28] Even if he dictated the book to someone else (which seems a likely possibility), Lomazzo's choice of words in the *Trattato* is of particular interest and constitutes one of the focal points of this monograph. I pay special attention to the terminology employed by Lomazzo to systematize the knowledge he possessed on the principal parts of painting, since his language in the treatise is revealing of the transformation of the practical knowledge of a painter into theoretical knowledge. One example, discussed below in ch. IV, is Lomazzo's account of the principal colors in the third book, *Del colore*, where he selects terms specifically used by artists – such as *pavonazzo* or *pallido* – to construct his complex system of colors. Moreover, as we shall see in ch. VI, his decision to employ "dialect" terms to describe the different parts of the human body is part and parcel of his aim to use the vocabulary of the artist's workshop to compile his theoretical system.

Since Lomazzo's academic activity was complementary to his literary achievements, it is given particular attention in the first part of this study in order to shed light on his milieu, as well as the people with whom he was in

26 Ciardi, *Scritti*, II, p. 588: "Usato un modo di dire ordinario e familiare, senz'alcun ornato, framettendovi anco molte parole meno approvate e che non si trovano usate da toscani." (Lomazzo, *Trattato*, p. 679).

27 Ciardi, *Scritti*, II, p. 589: "Quanto alle parole meno approvate, elle sono così proprie di quest'arte, e per conseguenza così significanti appresso i pittori, che non si potevano in alcun modo tralasciare, volendo essere inteso." (Lomazzo, *Trattato*, p. 680).

28 Jean Julia Chai, *Idea del Tempio della Pittura* (University Park, Pennsylvania: The Pennsylvania State University Press, 2013), introduction.

contact. He was a prominent member of the *Blenio* academy, a burlesque literary institution which included artists and artisans among its membership, the heterogeneity of which has not been stressed enough.[29] In the list of the academy's members we find painters such as Ottavio Semino, Camillo Landriani (known as "Il Duchino") and Aurelio Luini.[30] According to Lomazzo, Aurelio possessed a little book (*libricciuolo*) full of drawings in Leonardo's style; this is an important hint as to the ways in which painters shared knowledge and visual material in their gatherings, as we shall see in ch. VI of the monograph, on the study of anatomy.[31] The *Accademia* included not only painters but also craftsmen who worked in the realm of the so-called "decorative arts": Annibale Fontana, Scipione Delfinone, Francesco Brambilla and Francesco Tortorini. They were central to the creative economy of Milan, internationally renowned at the time for the production of luxury goods. During those years, the city hosted painters from other regions and from outside Italy, engaged in carrying out works for the dioceses of Carlo Borromeo, and they had the chance to interact with local painters, creating a dynamic and probably quite competitive atmosphere.[32] Treating Lomazzo's own artistic production has always been a challenge, and his pictorial works (at least, the ones we know about) are taken into account as examples of the topics discussed in each chapter (for instance, the frescoes he painted in Milan and Piacenza are used to discuss his notion of foreshortening).[33]

The overarching aim of my research has been to reconsider Giovanni Paolo Lomazzo, a central figure in art history, by placing his achievements in the wider context of the history of knowledge. This monograph therefore combines an art historical approach with the history of reading and of books, in order to achieve a deeper understanding of the making of Lomazzo's books

29 The most complete study on the academy's activity is Dante Isella's edition of *Rabisch* (Turin: Einaudi, 1993).

30 For recent research on Luini, see the exhibition catalogue: Giovanni Agosti, Iacopo Stoppa and Rossana Sacchi (eds.), *Bernardino Luini e i suoi Figli* (Milan: Officina Libraria, 2014).

31 Lomazzo, *Idea*, ch. XXVI, translated in Chai 2013, p. 89: "These figures are scattered all over the world, except for those drawn in sanguine which the Milanese painter Aurelio Luini has in his possession. Among these drawings, there are some figures who laugh so heartily, by the power of an immense art, that nature herself can barely equal them."

32 The exhibition on Bernardino Luini in Palazzo Reale also added new documents related to Aurelio Luini, Bernardino's son, which testify to his prolific activity as painter and to his exchanges with foreign painters during his working years. See Agosti et al. (eds.), *Bernardino Luini e i suoi Figli*, especially the *Regesto dei Documenti*, pp. 386–389.

33 See ch. V below.

and the nature of his sources. Lorraine Daston has highlighted: "The implications of the history of scientific reading for other, more familiar forms of scientific practice, such as observation, but also for what might be called cognitive practices."[34] If we replace the word "scientific" with "artistic" or, more appropriately, "artisanal", it becomes clear that Lomazzo can be seen from a different perspective, as an example of the tendency to collect and publish material from the realm of "pictorial knowledge", translating it into written form. This approach is supported by recent scholarship exploring the transmission of artisanal practices from the workshop to paper (and, in this case, to a treatise). At the same time, it focuses on the making and transmission of a dedicated artistic vocabulary and terminology.[35]

My decision not to take into account all of the "theoretical" books in the *Trattato dell'Arte* – leaving out Lomazzo's discussion on proportion, on movement and to some extent on light – was made for practical reasons. By reducing the scope of this study, I have been able to concentrate on providing a close analysis of three major aspects of the treatise: color, perspective and anatomy. My purpose is to connect these three topics to his milieu, which is described in Part 1: "Lomazzo and Milan." By investigating his heterogeneous connections, not only with artists but also with the local *intelligenzia*, I was able to uncover information on a variety of topics (his direct and indirect relationships, for example, with anatomists in Milan such as the Carcano brothers or with the mathematician Giovanni Battista Benedetti).[36] His connections to the aristocratic patron of the academy, Pirro Visconti Borromeo, are explored in ch. III, showing how Borromeo's interest in the wonders of nature had a concrete

34 Lorraine Daston, *Things that Talk: Object Lesson from Art and Science* (New York: Zone Books, 2004), p. 444.

35 I am referring to the work of Pamela Smith, Pamela O. Long, Paula Findlen and Sven Dupré, among others. See, e.g., Benjamin Schmidt and Pamela Smith (eds.), *Making Knowledge in Early Modern Europe: Practices, Objects, and Texts, 1400–1800* (Chicago: University of Chicago Press, 2008); Pamela Smith, "The History of Science as a Cultural History of the Material World" in *Cultural Histories of the Material World*, edited by Peter N. Miller (Ann Arbor: University of Michigan Press, 2013), pp. 210–225; Pamela O. Long, *Openness, Secrecy and Authorship* (Baltimore: The Johns Hopkins University Press, 2001) and Pamela O. Long, *Artisans/Practitioners and the Rise of the New Science* (Corvallis: Oregon State University Press, 2011); Paula Findlen and Pamela O. Long (eds.), *Merchants and Marvels: Commerce, Science and Art in Early Modern Europe* (London and New York: Routlege, 2002); Mark Clarke, Bert De Munck, and Sven Dupré (eds.), *Transmission of Artists' Knowledge* (Brussels: Koninklijke Vlaamse Academie van Belgie voor Wetenschappen en Kunsten, 2011); Harold Cook and Sven Dupré (eds.), *Translating Knowledge in the Early Modern Low Countries* (Zurich and Münster: LIT Verlag, 2012).

36 See chs. IV and VI below.

impact on the project to renovate his *Villa* in Lainate, outside Milan,[37] for which Lomazzo was reputedly an advisor and consultant. As Michael Baxandall stated: "The gem even of Vasari's great *Lives of the Artists* lay in dinner conversation at Cardinal Farnese's, as he says himself, and the most vigorous roots of his books run down to workshop argument, two or three centuries of it."[38] Were the topics that Lomazzo chose to include in his books actually discussed by the artists and men of letters of the time? What were the dynamics of these conversations in which they interacted and categorized pictorial knowledge? In this study I attempt to answer these questions by making an effort both to identify the people to whom Lomazzo was talking and to specify the subjects about which they were conversing.

---

37 See ch. III.

38 M. Baxandall, *Patterns of Intentions* (New Haven and London: Yale University Press, 1985), p. 137.

PART 1

Lomazzo and Milan

CHAPTER 1

The Artist and the Traveller

I Vari Caprizzi che Strani mi Venian: Lomazzo's Education and Background

As we read in the autobiography written at the end of the *Rime* (1587), Lomazzo was born in Milan on the 26th of April 1538, at 5 p.m., a day astrologically "dedicated to Venus":

> In the year 1538, on the 26th of April, on a day sacred to Cytherea, I was born in Milan, at five o' clock. Growing up I always had my mind focused on drawing, so I was sent to a master when I was ten years old, who taught me to write and to count, to handle books, and then to draw. With these beginnings I started to paint under a follower of Gaudenzio Ferrari, a respectable painter named Giovanni Battista della Cerva. He encouraged me, seeing that I was prepared for success in this art. So at that time I executed various works: paintings, caprices, storie, friezes, grotesques and various decorations with cartouches, trophies, landscapes and fruits, which I painted in the three styles.[1]

1 Giovanni Paolo Lomazzo, *Rime*, pp. 529–530:
"De la salute nostra gl'anni mille
Cinquecento trent'otto, et vintisei
Giorni d'Aprile, et ore dieci sette
Correvano; quand'io nacqui in Milano,
In un giorno dicato a Citerea.
Così crescendo havea sempre la mente
Intenta al disegnar; e intanto poi
A mastro me n'andai fino a dieci anni
Dove apprendei a legger et contare
E' l maneggiar de i libri, et poi disegno.
Con tanti principi al pingere mi dei
Sotto un discepol del morto Gaudenzio
Ferrari, che fu già degno pittore
Nomato Gian Battista della Cerva.
Il qual mi spinse innanzi contemplando
Quel ch'io era pronto a far nell' arte sola:
Però in quei tempi feci diverse opre,
Sì come quadri, bizarrie, historie,
Fregi, grotteschi, et partimenti vari
Con carrozzi, trofei, paesi e frutti
Quai variando in le tre sorti pinsi."

 | DOI 10.1163/9789004330269_003

Lomazzo was the son of Giovanni Antonio Lomazzo and Lucia Mozzanica. Although we do not know his father's profession, he probably came from the Milanese minor nobility, and it is also probable that his father had an interest (whether it was linked to his profession is hard to say) in figurative and decorative arts, since three of his four sons became painters (Giovanni Angelo, Pomponio, Giulio Cesare) and the fourth (Girolamo) became an embroiderer. The Lomazzo family lived in Milan in the parish of Santa Maria Beltrade. Antonio Lomazzo had close relationships with different personalities belonging to the Milanese nobility such as Melchiorre Mazzenta, Prospero Visconti and a nobleman from Piacenza, Alessandro Castiglioni. At the same time he was acquainted with many families of artists and artisans belonging to the same parish, especially with the *botteghe* of the Delfinone family (one of the most successful embroiderers at the time) and the Figino family, of which Giovanni Ambrogio became the most renowned of Lomazzo's pupils. Among the relationships with artisans, the one with Scipione Delfinone is particularly worth mentioning. Scipione, who became member of the Academy of Blenio that had Lomazzo as its chief for many years, took Girolamo Lomazzo as a pupil in his *bottega*.

Not much is known about Giovanni Paolo's education as a young boy (see fig. 1.1). He certainly went to a teacher to learn how to write and read, to count, and, as he says, to "handle books" (*maneggiare i libri*).[2] He also went to a music teacher, named Gian Michele Gerbo. His main teacher in painting was Giovanni Battista della Cerva, a pupil and colleague of Gaudenzio Ferrari (1478–1546). This fact is confirmed by a document dated February 1552.[3] Despite this document, there is a passage of the Treatise in which Lomazzo defines Gaudenzio Ferrari, and not Della Cerva, as his "old master."[4] However, one cannot ignore the fact that Lomazzo's birth date could be traced only ten years before the death of Gaudenzio and the two probably did not come into contact directly. This inconsistency has been used to "claim" that Lomazzo's information on himself is unreliable, but I tend to believe that the artist started his period of apprenticeship before Della Cerva, perhaps in his own family of artists, when he was ten years old, as he states in the beginning of his autobiography. Then,

---

2 Lomazzo, *Rime*, p. 529.

3 In Giulio Bora, Manuela Kahn-Rossi, Francesco Porzio (eds.), *Rabisch, Il Grottesco nell'Arte del Cinquecento* (Milan: Skira, 1998), p. 329.

4 Ciardi, *Scritti sulle Arti*, II, p. 101: "Non fu secondo il mio vecchio precettore Gaudenzio, non solamente saggio pittore, come ho detto altrove, ma profondissimo filosofo e matematico." (Lomazzo, *Trattato*, p. 112).

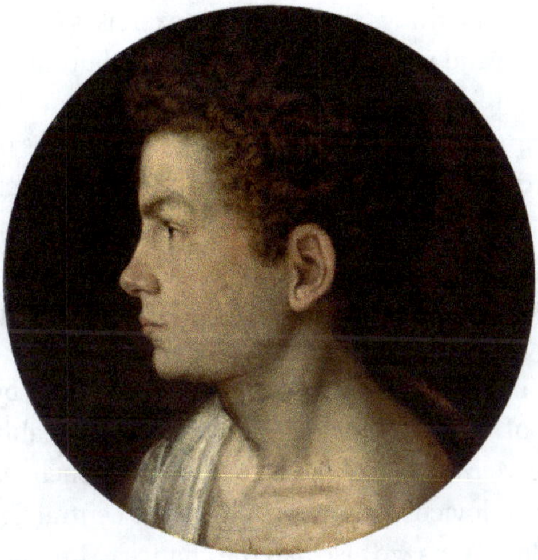

FIGURE 1.1
Giovanni Paolo Lomazzo, Self portrait as a young boy, *c. 1570, Vienna, Kunsthistorisches Museum*

in 1552 as the document attests, that is when he was fourteen years old, he became a pupil of Della Cerva.

On his first activity as a painter, we read that in the *studio* of Della Cerva he was trained to paint different subjects: "*storie*, grotesque and peculiar figures, portraits and all sorts of figures."[5] He was most likely helping his master to decorate and complete the ornaments of paintings, a crucial part of the training to develop his own style.

The apprenticeship in the studio of the master lasted until 1559, when Lomazzo was twenty-one years old. In this year a legal dispute arose between master and pupil, of which we know little, since there is only one sentence documenting this quarrel. The *liberatio* of Giovanni Paolo from Della Cerva occurred because the latter somehow did not respect the pact stipulated seven years before.[6] One year later, Lomazzo, at twenty-two years of age, signed the document of *emancipatio* to free himself from the parental tutorship.[7] The first document which attests the activity of the artist as a master is dated January 1564: it constitutes a pact between Ambrogio Figino's father, Vincenzo Figino, and Giovanni Paolo. At the time he became a pupil Figino was eleven years old:

> They gather together here Vincenzo father of Ambrogio his son (…) who must go to the house of the aforementioned Giovanni Paolo, in order to

5 Lomazzo, *Rime*, p. 529: "Sì come quadri, bizzarrie, historie, fregi, grotteschi e partimenti vari con cartozzi trofei paesi e frutti."

6 Bora, Kahn-Rossi and Porzio (eds.), *Rabisch*, p. 329.

7 Ibidem.

> learn the art of painting and all the things that Lomazzo exercises (*in his workshop*). And he will practice and work to the advantage and utility of Lomazzo, and he will obey him for the next six years, starting from this January, both during the day and the night, according to the costumes of this art. And the aforementioned Giovanni Paolo will have to teach Ambrogio this art.[8]

Similar documents have been found for the pupils Girolamo Ciocca (May 1569) and Pietro Martire Ottolini (July 1570).[9]

As a master in his workshop, Lomazzo had paintings commissioned by many local noblemen and men of letters whom he knew most probably through his participation in the *Accademia de la Val di Blenio*, such as Sigismondo Foliani and the writer Lodovico Gandini.[10] These were central figures in the local *intelligenzia* of Milan. Lomazzo also had the chance to interact with many of the artists and artisans that lived near his parish, since, as we shall see, they carried out various works for the dioceses of the city, creating a dynamic and rather competitive atmosphere, which the painter mentions in his writings more than once.[11]

---

8 Bora, *Rabisch*, pp. 329–330: "In primis convenerunt quod dictus Vincentius teneatur facere et curare cum effectu quod dictus Ambrosius eius filius (…) tenetur ire ad domus habitationis dicti Domini Jo. Pauli ad adiscendum artem pingendi et de omn. exercitio quod exercetur per dictum de Lomatio, et se et persona exercere et laborare ad commodum et utilitatem dicti de Lomatio et ei praestare obedientiam et hoc per annos sex proximos futuros, incepturos in Calendis Januarij presentis, et de die et de nocte secundum stilum et consuetudinem dicte artis. Et dictus dominus Jo. Paulus teneatur eum Ambrosium instruere in dictam artem."

9 Bora, *Rabisch*, p. 331.

10 See Lomazzo, *Rime*, p. 530: "E ne' ritratti ancor io posi il piede/Di piccioli, et di grandi." Lomazzo lists many portraits that he made at young age (pp. 531–533). These portraits are now lost or not identified.

11 An issue in dealing with artists and painters from the North of Italy has always been to find reliable sources. There is, unfortunately, no local biographer like Vasari for Florence. Although it is recognized that the latter cannot be considered a fully reliable source, his account on Northern Italian artists in the second edition of his Lives of the Artists is of some interest, especially regarding the lives of Benvenuto Garofalo and Girolamo da Carpi. However, Vasari's information is somewhat random and anecdotal as it derives, for the most part, from the travel to the North of Italy made by the author. See Giorgio Vasari, *Vite de' più Eccellenti Pittori, Scultori e Architetti Moderni* (Florence 1568), ed. by Gaetano Milanesi (Florence: Sansoni 1881), 9 vols., vol. 6, pp. 457–520, p. 457: "In questa parte delle Vite, che noi ora scriviamo, si farà brievemente un raccolto di tutti i migliori e più eccellenti pittori, scultori et architetti che sono stati a' tempi nostri in Lombardia, dopo il

Little survives regarding his activity as a painter, and many of the paintings and portraits that the artist references in his autobiography are now lost or not acknowledged. A much needed study has fairly recently been done by Mauro Pavesi who, in his dissertation "Giovanni Paolo Lomazzo Pittore Milanese, 1538–1592", took into account the activity of the author as an artist, partly filling the gap of his pictorial production and especially re-evaluating his status as a renowned portraitist in his own time.[12] Declarations of esteem by his contemporaries are not limited to the painter's pictorial production (which presumably stopped around 1571, the year in which he was struck by the eye-illness). Local men of letters and poets knew and praised Lomazzo's literary achievements with dedications and sonnets on different occasions. The literary figure Lodovico Gandini, one of the members of the Blenio Academy, dedicated three sonnets to the artist, which are published at the beginning of the *Trattato*, in the customary celebration of the book's author. One in particular, *I pittori havean luci* (painters had lights), underlines the alleged didactic role which the work of Lomazzo had for painters at the time: "Painters had lights/But they did not paint with lights/Now not only they will have lights/But they will also paint with light/Who gives the light?/Your enlightening work/Highest enlightened painter/Who has no light."[13] This poem is followed by two sonnets by Baldini and by Foliani, two other members of the academy. These are written in Latin, probably to testify to the authors' culture and to distinguish themselves from other writers.[14] In addition to this, the literate Goselini (whose name in the Blenio Academy was *Goselin Slurigliagn*), who had commissioned the painting *Cristo nel Giardino degli Ulivi* from Lomazzo (now in the Brera), referred, in his *Rime di Diversi Celebri Poeti dell'Età Nostra,* to Lomazzo as a poet worthy of praise, who talked and taught about art. If he previously had the brush as his

Mantegna, il Costa, Boccaccino da Cremona et il Francia bolognese, non potendo fare la vita di ciascuno in particolare, e parendomi a bastanza raccontare l'opere loro; la qual cosa io non mi sarei messo a fare, né a dar di quelle giudizio se io non l'avessi prima vedute. E perchè dall'anno 1542 insino a questo presente 1566 io non aveva, come già feci, scorsa quasi tutta l'Italia, né veduto le dette et altre opere, che in questo spazio di ventiquattro anni sono molto cresciute, io ho voluto, essendo quasi al fine di questa mia fatica, prima che io le scriva, vederle e con l'occhio farne giudizio."

12 Mauro Pavesi, *Giovanni Paolo Lomazzo Pittore Milanese 1538–1592* (Milan: Università Cattolica del Sacro Cuore, Doctoral thesis, 2006–2007).

13 Lomazzo, *Trattato,* folio 2: "I pittori havean luci/Ma non pingean con luce/Hor non solo havran luci/Ma pingeran con luce./Chi porge la luce?/L'opra lucente tua, privo di luci/Altissimo Pittor colmo di luce."

14 Lomazzo, *Trattato,* folio 3.

instrument, "now he has the pen to give prestige to the world."[15] He alludes of course, as Gandini also did, to the blindness that struck the author at the age of thirty-three, cutting short his career as a painter.[16]

Respect for and interest in the literary works of Lomazzo can be traced in various sonnets as well. Two sonnets at the beginning of the book *Rabisch* seem particularly worth noting, especially considering that they were written by fellow artists, namely Ambrogio Brambilla (the founder of the *Accademia di Blenio*), and Girolamo Maderno.[17] The former, described by Lomazzo as "painter, carver, propounder and inventor of this valley", was the first to hold the position of great counsellor, one that Lomazzo would later hold (by then the name of the role changed into *abate*).[18] In his sonnet "In honor of Zarvagna and the valley", Brambilla states that:

> You know, fellow Zarvagna (*the academic name of Lomazzo*), among us we can admit, without pretending, that nobody's as worthy as you. Moreover, I will tell you this: in writing grotesques, all the poets of the valley are worthless compared to you. And nowadays you sing the praises

---

15 Giuliano Goselini, *Rime di Diversi Celebri Poeti dell'Età Nostra* (Milan: Comin Ventura e Compagni, 1587), p. 239: "Chi può cieco chiamar un ch'a l'oblio/Tolsi d'eterna notte? Un, ch'Argo al Cielo/Vola? Di cui non hebbe il mortal velo/O madre antica, figlio unqua sì pio?/Cieca, e muta giaceasi, ed aspro, e rio/Sentia già del suo fine il freddo gelo,/Del pinger l'arte; e'l costui caldo zelo/Gli occhi, e le labbra a lei languente aprio:/C'hor parla, e mira, e per lui vive, e'nsegna:/Et ei, fatto per lei maestro egregio,/La sua pietà con doppia gloria illustra./Già del pennello, hor de la penna ha'l pregio:/Ne de la cetra sua quel Dio lo sdegna,/Che pingendo, e cantando il mondo lustra." On Giuliano Goselini, an understudied poet and intellectual figure in Milan, see the recent article by Luisella Gioachino, "La Lode e la Morte: Giuliano Goselini Poeta Funebre della Milano del Secondo Cinquecento," *Allegoria*, 2009/2010, *XXVI*: pp. 102–130.

16 Lomazzo, *Rime*, pp. 538–539.

17 For information on Ambrogio Brambilla see Isella, *Rabisch*, pp. 332–334. On Girolamo Maderno unfortunately not much is known. Ciardi in his *Scritti sulle Arti*, I, XIX, labels him as an architect, maybe one of the brothers of the better-known Milanese architect Carlo Maderno, but without giving detailed references to his work. Isella instead considers him a painter (Isella, pp. 363–364), giving as a proof a sonnet by Lomazzo, who listed Maderno among the painters in the above-mentioned dream (Lomazzo, *Rime*, p. 134: "E di que' tra i pittori/Era Carlo Cremasco col Maderno").

18 Isella, *Rabisch*, p. 83: "Ar signó Bosign Brambilla penció, intaglió, gettó e trovaglió de costa val e persció è ciamad or compà Borgnign Gran scanscieré de Bregn."

of Bacchus and of the porters from the Blenio valley so well that the whole world is amazed.[19]

The sonnet by Maderno (known in the academy under the name of *Compà Ciabócch*), not only praises Lomazzo's skills as a writer, but, more specifically, proclaims the usefulness of his books for those who love drawing:

> Who has ever seen a mind comparable to Zarvagna's, our *Abate*, who, when he was able to see, always competed with good painters and, when he lost his eyesight, started to write carefully on his profession, in such a way as to enlighten all the people who love drawing? A work so valuable that everybody was astonished, thinking how could a blind man have published such a precious book? Indeed, was this not a useful mirror for the ignorant, to find that beauty which was hidden is now revealed? Did he not accomplish more than Dante, in finding such a witty invention as the Arabesques, which he published, and also (*in writing*) this book which he put together, full of the habits of the porters from the Blenio valley?[20]

Apart from the pretentious comparison with Dante, this sonnet implies that Lomazzo's books were actually taken into consideration and discussed by artists, at least by those in close contact with him. It is reasonable to suppose that during his period as *abate* of the academy, that is from 1568, discussions on the status of the arts and artists in Milan were common among artisans and men

19 Isella, *Rabisch*, pp. 9–10: "Te sé compà Zarvagna che tra nugn/Pòm divisò senza simulaçiogn:/Che varen più de tì no ghè nissugn./Da pu te vugl' anch dì costa resogn,/Ch'in fà grotisch hign al par tò minciogn/Tucc i poglita ed qua 's vuglia valogn./E mò dor nòst Baccogn/E di fachign dra val de Bregn te cant/Tant begn ch'o 's maraveglia or mond tutt quant." It is worth noting Lomazzo's reply in the second part of the book (Isella, pp. 85–6), in which, out of false modesty, he states that: "You shouldn't in your discourses describe me as such a learned man." ("Ma ti o 'n disiss sciert sciá cogl' tò resogn/Fam insci dott").

20 Isella, *Rabisch*, pp. 11–12: "Chi ha magl' vist nè sentud on tal insciegn/Come quòl dor Zarvagna nòst Nabad,/Che quand or ghe vedeva è sempro stad/Cogl' bon mena penigl' a bòtta e a segn;/E quand o r'ebb dra lus pers o sostegn/Or se miss a scriv sciò dor gran bel grad/Dra soa profesigliogn che inluminad/O r'ha tucc quògl' ch'o amen or dessegn?/Còsa che ogni vivent se'n stà pensos/A pensà come on òrb abbia facc tant/Da mandà fu on librasc sì preçiglios./Mò n'el lú stacc on specc agl'ignorant/A' vè trovò quòl bell che s'era ascos/E a' vell spantegò fura dar sò mant?/Nè hal facc piú ca Dant/A'vè si bella invençiglion trovad/Com' quòlla di Grotisch, ch'a r'ha mandad/In stampa, nè è bastad,/E quòst ch'a r'ha compost, on liber piegn/Di costum di fachign dra val de Bregn?"

of letters, who interacted in informal (and public) environments such as the Blenio academy.

Another indication of esteem for the artist's literary production comes from the writer on art Gregorio Comanini, author of the *Figino*, which was published two years after the *Rabisch*. He commends the work of his Milanese colleague in a sonnet at the beginning of the *Idea del Tempio della Pittura*, stating that Lomazzo built up a "golden temple which has, instead of arches, writings and solid pages, better fit than marble to endure the hard tooth of time, pages in which learned decisions, clear voices, gentle sensations, and deep secrets are explained."[21]

Of course this and the other *laudatio* follow the lines of the praises dedicated to the author at the beginning of the book, but aside from acknowledging their rhetorical function, they are interesting to mention in order to see who was part of Lomazzo's network in Milan beyond painters. Further indication of the popularity of Lomazzo's writings is the fact that the *Treatise* was translated into English quite early on by the Oxford scholar Richard Haydocke in 1598, only fourteen years after the first edition.[22] All these declarations of esteem testify to Lomazzo's popularity as a writer on art even beyond his literary circle. Furthermore, they position him as a respected personality in the city of Milan.

The Author's Travels

Lomazzo is not very eloquent with regard to his travels through Italy at a young age (roughly between 1559 and 1571), but from his autobiography we can attempt to understand his peregrinations during his youth. We have an explicit reference to his travel to Rome in the *Rime*, when he states that: "Desiring to obtain virtue I went to Rome, to see the paintings and the *anticaglie*, of which

21 Gregorio Comanini, *Il Figino ovvero del Fine della Pittura* (Mantua: Francesco Osanna 1591). For the sonnet see Klein, *Idea*, I, p. 9: "E pur un aureo tempio a tua memoria/Formi, che 'n vece d'archi ha prose, e carte/Carte via più che marmi a'i duri denti/Del tempo salde, ove son dotti intagli,/Chiare voci, almi sensi, alti secreti."

22 Richard Haydocke, *A Tracte Containing the Art of Curiose Painting* (Amsterdam and New York: Da Capo Press, 1969). This translation is a most interesting source that I do not analyse in detail here, though it will be mention in different occasions in the study. The English translation of the Treatise opens up new lines of research concerning the fortune of the book outside of Italy and the rising of a discourse on painting in England at the end of the sixteenth century. A critical edition by Dr. Alexander Marr should appear soon.

my eyes and my mind were never tired."[23] He declares that he saw all the "antiquities", which probably refers to a visit to the *domus aurea* in order to see the grotesques, but it could also refer more generally to his visit to the city's monuments. The artist then refers to his travels through the north of Italy and he lists the pictorial works that he achieved in various cities: he mentions the fresco of the *Cena Quadragesimale* for the church of Sant'Agostino in Piacenza (which is now lost and of which only a preparatory drawing survives, see figure 5.13) and his works in Lodi, not far from Milan, for the church of San Romano.[24] Of his journeys, two alleged travels to Turin and to the Netherlands, likely made before 1570, are interesting to investigate further in order to understand the artist's connections beyond his native Milan.

It is known that Lomazzo dedicates the Treatise to Don Carlo Emanuele of Savoy, duke of Turin, as he also reminds the readers in his autobiography.[25] Although he does not specifically declare that he has been staying at the court of Turin, this dedication, along with some explicit references in his poems to people and men of letters present at the court at the time, leaves little doubt that the painter was familiar with and must have visited this city at some point.[26] In book II of the *Rime*, Lomazzo dedicates a poem to Ottavio Semino, colleague and member of Blenio under the name of "Compà Argh." He states that: "I saw in Turin Ottavio dining with many others at the *osteria Capello*."[27] Apart from this direct reference, one can find different poems dedicated to people serving at the court of Carlo Emanuele: for instance, to the artist Alessandro Ardente, who moved to Piemonte around 1572 and who was appointed as official sculptor under the father of Carlo Emanuele I, Prince Emanuele Filiberto.[28]

He was also in contact with mathematicians from this environment, such as Giovanni Battista Benedetti and Francesco Ottonai, to whom he dedicates two

---

23 Lomazzo, *Rime*, p. 533: "Di virtù poi bramoso andai a Roma/Per veder le pitture et anticaglie/Le qual mirar et osservar giammai/Non furon gl'occhi et la mia mente satia."

24 Lomazzo, *Rime*, pp. 534–535: "Quindi andai a Piacenza, et ivi fei/Nel rifetorio di Santo Agostino/La facciata con tal historia pinta (...) /In Lodi poi due tavole dipinsi/Di San Roman nel tempio."

25 Lomazzo, *Rime*, p. 540: "Le qual sacrai al sommo duca Carlo/Di Savoia, splendor, et chiaro raggio/D'ogni real eroica virtute."

26 On Lomazzo's travels to Turin and Antwerp, see Ciardi, *Scritti* I, VIII, footnote 9.

27 Lomazzo, *Rime*, p. 110.

28 Lomazzo, *Rime*, pp. 110–111.

poems in book III of the *Rime*.[29] In the nineteenth chapter of his *Idea* Lomazzo refers to both mathematicians as the authors necessary to read and study in order to understand geometry. As we shall see in the chapter dedicated to perspective, he was also familiar with Benedetti's works on optics and he probably knew him personally.[30]

Another indication that the author might have visited Turin is his knowledge of the political situation there. In a different poem Lomazzo mentions: "When Emanuele, Prince of Savoy, punished the blasphemy of his own people", probably referring to the repression against the Valdesi people around 1560–1561.[31] Why and for how long the artist decided to visit the court is difficult to establish, however, the dedication of the treatise to Carlo Emanuele, "patron of the arts and of the sciences", allow us to assume that Lomazzo must have felt a debt of gratitude for his time spent at the Turin court.

For what concerns the author's travel to the north of Europe, and especially to the Netherlands, we know that he dedicates a poem to Frans Floris de Vriendt (or Francesco Flor, as he calls him), in which he declares: "Among the most excellent painters I say that there is no one as learned and bizarre as Francesco Flor, whom I saw drinking and painting in Antwerp, when I was there ill."[32] He then mentions a certain "Lhemschercho" who painted Dutch women, which has been identified as the painter Maarten van Heemskerck.[33] As Ciardi notes, given the fact that Floris died in 1570, and if we trust Lomazzo's

---

29 Lomazzo, *Rime*, p. 158.

30 See ch. V.

31 Lomazzo, *Rime*, p. 240. Mentioned in Alessandra Ruffino (ed. by), *Rime* (Rome: Vecchiarelli Editore, 2006), p. 284, n271.

32 Lomazzo, *Rime*, p. 107:
"Tra i più eccellenti e gran pittore io affermo
Non haver visto il più bizzarro e dotto
Di quel ch'io vidi senza pagar scotto
Che fu Francesco Flor, non men ben fermo
Nel bere che nel pinger di buon trotto.
Il qual mi disse: hor beviam pien un gotto
Di vin che ci sarà nel pinger schermo.
Lhemschercho vidi ancor, che le sue Olande
donne dipinse con pastoso e secco
Pennel, col fil tempera d'una saga.
Pe'l qual i dotti cinti di ghirlande
Giurar per la persa anima d'un stecco,
Lasciar a la moglier portar la braga."

33 Rogier van Son, "Lomazzo, Lampsonius en de Noordelijke Kunst," *Nederlands Kunsthistorisch Jaarboek*, 1993, *44*:185–196, p. 187.

words, we can assume that he visited Antwerp before that date.[34] In his autobiography, however, there are no references to this trip, a fact that has been used as evidence that he did not actually make such a journey. Yet one should note that the author offers scant information also on other travels that he most likely did take, such as to Florence and to Naples. To support this argument, there is an interesting Latin poem dedicated to Lomazzo in the first edition of the *Trattato* from a certain "Guglielmi Huismanni Antwerpiensis", who praises the author's literary achievements that are, according to his words, made "in seven books dedicated to Carlo" and "after a terrible eye-illness took away the sight from you", confirming that the author of this poem was also aware of Lomazzo's illness and therefore was probably in contact with him personally.[35]

---

34 On Lomazzo's travels to the North of Europe see also Roberto Paolo Ciardi, *Ambrogio Figino* (Pisa: Marchi & Bertolli, 1968), pp. 16–17. He attributes the reason for these journeys to Lomazzo's desire to distance himself from the cultural hegemony of Rome and Florence. He argues that Lomazzo's acquaintances with artists from the Netherlands could also derive from the presence of different Dutch artists for the *Fabbrica del Duomo* at the time. On the topic see also M. Calvesi, "Contribution de Gian Paolo Lomazzo à la critique des Fiamminghi," *Les Arts Plastiques*, 1951, *II*:131–134. On the cultural relationships between Milan (and Northern Italy in general) and the Low Countries see for instance Ingrid Alexander-Skipnes (ed. by), *Cultural Exchange between the Low Countries and Italy (1400–1600)* (Turnhout: Brepols, 2007).

35 Lomazzo, *Trattato*, folio 10:
"In commendationem operis Io. Pauli Lomatij,
Carmen Guglielmi Huijsmanni Antwerpiensis:
Quidquid Apellea quondam fuit arte retrusum
Praeclaros potuit quod latuisse viros:
Quicquid et artifici penicillo pinxis Apelles,
Quod valuit nemo, dexteritate pari:
Hoc vigili studio septena volumina pandunt,
Nominu auspicio Carole celse tui.
Continet undosum mare nil, nil terra, polusq:
Quod tegit obliquis axis uterq; rotis,
Quod non egregio conatu expressit as unguem.
Naturam artifici concomitante manu.
Occulta tacito nil mens humana recessu,
Cuius non quaedam semian vultus habet.
Cedite Romani pictores, cedite Gracei,
Multiplici gratum nomine prodit opus.
Sed quo Paule modo, qua te celebrabo camoena?
Obstupet in dote nostra thalia tuas.
Lyncea te fecit natura, sed horridus orbum
Casus, natura haud, haud tibi casus obest.

Moreover, in the autobiography Lomazzo declares that he made a painting for Hans, the nephew of Dürer: "This was made for the nephew of the great Dürer", though unfortunately he does not give more details.[36] As we shall see in part II, Lomazzo's knowledge of Northern experts on perspective goes beyond Dürer. In the Treatise, he claims to know Northern authors who refined perspective methods, such as Hans Vredeman de Vries and Hans Lencker.[37] He mentions in passing that he "saw the instruments" of these authors, and "many other drawings made with the perspective of Lencker."[38] Although the artist does not elaborate on these references, these hints could be an indication not only that Lomazzo probably visited Antwerp personally, but that he was certainly interested in understanding the working methods used there. His knowledge of Northern art is further reflected in the landscape that he depicted for the fresco of the *Cena Quadragesimale*. Lomazzo's information on Northern art may have also derived indirectly, from other authors, as Van Son has argued (1993), especially concerning the Northern painters that he lists in different parts of his writings.[39] However, apart from the poem dedicated to Frans Floris, Lomazzo's

---

Lumine que lynceus quondam penetravit acuto,
Ingenij superas lumine Paule boni."

36 Lomazzo, *Rime*, p. 538: "In una icona fei, qual è a man destra/In san Giovan in conca, un Christo in croce;/Che parla con la madre, e v'è Giovanni/Con Madalena al pié, questa al nipote/Feci del gran Durero, et una altra anco."

37 See ch. V of the work.

38 Ciardi, *Scritti*, p. 277: "Oltre a quello in altri modi si possono crescere dalle piante i corpi humani, come per forza di numeri col velo di Leon Battista Alberto, col telaro e la graticola di Alberto Durero e di Giovanni di Frisia di Graminge, i quali istromenti io ho veduti, insieme con molte altre figure disegnare da molti, con la prospettiva di Gio. Lenclaer." (Lomazzo, *Trattato*, p. 320).

39 Van Son refers especially to Vasari's *Vite* and Domenicus Lampsonius's *Pictorum Aliquot Celebrium Germaniae Inferioris Effigies*, published in 1572, as sources for Lomazzo's knowledge of the art from the North. These sources may as well be the ones that the author utilizes. However, Van Son never takes into account Lomazzo's passage on the scholars of perspective, and he speculates that the artist derived his knowledge of all the Northern artists whom he quotes from his alleged collection of prints. He interprets Lomazzo's possession of "four thousand papers and more" as such. Though Lomazzo must have seen many prints by Northern artists, especially, as he notes, in the house of the antique dealer Calistano, it is doubtful that he possessed such as huge collection himself. See Ciardi, *Scritti*, I, p. 174, n12 and Van Son, "Lomazzo, Lampsonius en de Noordelijke Kunst", p. 188: "Naast de werken die Lomazzo bij zijn vrienden en Calestano kon zien, fungeerde ook zijn eigen prentencollectie als bron voor zijn kennis over de noordelijke kunst. Zijn collectie grafiek moet indrukwekkend geweest zijn. In het *Trattato* vermeldt ihj dat zijn verzameling ongeveer vierduizend prenten omvatte." On the poem dedicated

list of the works of Northern *studiosi* on perspective, including Lencker and De Vries, is written in first person. This is especially relevant when he declares to have "personally seen" their instruments. In this important passage, which we will analyze in more detail in the chapter dedicated to perspective, the author tries to give the readers an exhaustive list of books and methods from which the artist should learn for the sake of his education, both literary and practical.

The Education of Artists

Lomazzo's opinion on the education of the artist is clear throughout his writings: one cannot be a good artist without both theoretical and practical learning. Rules are fundamental for painters, who cannot learn the profession properly by other means. For this reason, he declares both in the *Trattato* and in the *Idea* that he intends for his literary works to help artists in their profession. In the *proemio* of the former, he states: "I, who desire most the profit and the usefulness of those who start to learn this art (*of painting*), decided to add a sixth book, in which I will treat in a practical way the topics I taught theoretically in the other five books."[40] He also adds, in the dedication of the *Idea*: "Moreover, having in mind the benefit that could follow, I have demonstrated to others how to imitate nature readily on a flat plane, of which consists the art of painting. This is something that few can understand without the light of rules."[41]

Lomazzo feels the need to achieve (and to suggest to painters to achieve) a solid theoretical education through books. In a passage of *Della Forma delle Muse* (1591), he states that in starting to study the art of painting he took into consideration all sorts of writings that he thought could have been useful to

to Frans Floris, Van Son argues that Lomazzo may not have been the author of it, although he does not provide any alternative suggestion of autorship. I wish to thank Dr. Sietske Fransen for helping me with the translation of Van Son's article.

40 Ciardi, *Scritti*, II, p. 23: "Io che sommamente desidero il profitto e l'utilità eziando di quelli che cominciano a imparare quest'arte, ho voluto aggiungere un sesto libro nel quale tratterò praticamente quello che ne i cinque libri si insegna teoricamente." (Lomazzo, *Trattato*, p. 15).

41 Ciardi, *Scritti*, I, p. 244: "Et in oltre pensando al giovamento che potea seguire, co'l dimostrare altrui la via spedita e piana d'imitare e come emular la natura, in che consiste tutta l'arte de la pittura. Cosa che da pochi senza il lume delle regole e dei precetti può essere intesa."

arricchir l'inventione del pittore ("to enrich the painter's invention"), consequently gathering the information in this "little book" which he published:

> I acknowledged this issue (*the ignorance of painters in representing the true form of the Muses*) from my first years of apprenticeship in painting, while I was reading all those pages by writers which I believed could be useful to enrich the painter's invention, observing many things on the nature and the condition of Muses, which I decided to gather in this little book, to publish them in order to be useful for painters and sculptors to learn the true form in which the Muses should be represented.[42]

Here his didactic aim and intended audience could not be clearer. His attitude towards "practical painters" who do not consider theoretical notions necessary for their work is very critical. He labels as ignorant those artists –so numerous compared to the learned ones– "who suffocate and infect the whole world, with the vagueness of their mere practice."[43]

42 Ciardi, *Scritti*, II, p. 595: "A questo avendo io, infin da i primi anni che mi diedi allo studio della pittura avvertito, mentre ch'andava rivolgendo le carte di tutti quei scrittori onde giudicava potersi arricchir l'inventione del pittore, molte cose intorno alla natura e condizion delle Muse osservai, le quali ora m'è parso di raccogliere in questo picciol libro, con quell'ordine col quale furon da me di tempo in tempo osservate, e divolgarle accioché di qui possano i pittori e scoltori apprendere la vera forma nella qual le Muse debbano rappresentarsi." Whether these *carte* are the same as those mentioned in his autobiography, of which he claims to possess four thousand, cannot be proved for certain.

43 Ciardi, *Scritti*, I, p. 330: "Che tutto il mondo ammorbano e soffocano con la vaghezza della pura prattica loro." (*Idea*, 1590, ch. XXXI). In another poem dedicated to the painter Paolo Camillo Landriani, named il Duchino, we can see how this time Lomazzo uses the term "practical painter" in a positive meaning, defining the colleague: "The most practical painter of our land, who shows to everybody what he is capable of, in *storie, fregi*, and in portraying with brushes people on the spot (…) He is friend of each manner of painting and working, such as frescoes, oil, gouache, and in scribbling on walls and doors with different techniques. (…) He is the honor of painting nowadays." In this case the word *pratico* (or, in the *Bregnese* dialect, *pratich*) implies his skills as a remarkably versatile artist, who masters several techniques and shows to the world (which is, the Milanese public) all his talents. (Isella, *Rabisch, pp.* 150–151: "Ar penchió Paol Camil Landrigliagn dicch or Ducchign et in dra Vall o r'è or Compà Squarta Maglia./Or piú pratich penció dor nòst paglis,/E ch'a ognugn fa veghè tutt or sò fà/In istorigl, in fris, e in dor retrà/Ra scient cogl' penigl' a r'improvis/O r'è quòl Landrigliagn che a facc fu amis/De tucc I mus del pensc e lavorà/Come in fresch, uiro e a squazz e in spegascià/Su per ancònn muragl' in tutt i guis./Ar pòrta de òr, tucc i bottogn taccad,/Agl' pagn de seglia, e de velud, e s'è/R'onò

In this respect, the *Blenio* Academy of which he was part, as we shall see, may be considered a disguised attempt by Milanese painters and artisans to join forces with local men of letters in order to fight against the predominance of artists ignorant of "good letters" (*buone lettere*), and in order to create a new "ideal type" of painter capable of discoursing on various topics and expressing himself poetically.[44] If Lomazzo shows contempt towards ignorant painters, it should be noted that he is equally scathing about painters who behave as courtiers, "who despise being called painters, and follow the habits of gentlemen and knights, caring only about gracefulness and fancy clothes. For these reasons they deserve nothing but to be pointed at and to be disdained."[45] This judgment fits perfectly with the unembellished, straight-talking attitude that the author and his fellow artists adopted in the *Accademia di Blenio*, as we shall see in the next chapter, which was demonstrated by the fictitious and ironical model of the unrefined porter (*facchino*).

Lomazzo, Astrology and Alchemy: Some Hints

During his youth, Lomazzo had dealings with and admired the work of different learned men of his entourage, such as Bernardino Baldini, who wrote two poems to praise the painter at the beginning of the Treatise, Guido Mazzenta, whom Lomazzo mentions more than once in his writings, and the physician and astrologer Girolamo Cardano.[46] In all probability, the artist not only knew the latter personally, but he also had knowledge of some of his books. Among the poems in the *Rime*, many are dedicated to learned men and two are dedicated to the "doctor and mathematician", as Lomazzo calls him.[47] Moreover, in

de ra penciura á quòst etad./Costu r'è on bell compagn, e spert o r'eè in lengua ed Bregn òd mud che solevad/La ra Fam fin dessora tucc i sciè.")

44 For Lomazzo's intention of presenting in his books a new "canon of the good artist," see the general conclusions.

45 Ciardi, *Scritti*, I, p. 334: "Sdegnando in un certo modo d'esser chiamati pittori, e seguitando le pratiche di signori e cavaglieri, attenti solamente a gentilezze, garbi e costumi. Onde altro non n'acquistano che esser mostrati a ditto e scherniti." (Lomazzo, *Idea*, ch. XXXI).

46 For the relationship between Lomazzo and Cardano, see B. Tramelli, "Due Poesie del Pittore Giovanni Paolo Lomazzo, Ammiratore di Cardano," *Bruniana et Campanelliana*, 2010, *XVI.II*:573–576.

47 Lomazzo, *Rime*, pp. 156–157:
"*Di Girolamo Cardano, Medico e Matematico*
Più non potea salir la fama vostra
Che per quei studi in qual sí dotto sete,

his autobiography the painter remembers how Girolamo had foreseen the blindness that struck him in 1572: "For a grave illness my eyes closed/And I lost the beloved light/And then I was beside myself/As the great Cardano predicted/A fine mathematician and doctor/Whom I portrayed with his sphere and his books."[48] Other poems attest to Lomazzo's interests in astrology and magic, such as the one dedicated to the astrologer Girolamo Vicenza: "All the astrologers gathering together stated that the worst evil is to be abandoned by

---

E pur potrà più ancor, se voi volete
Seguir quell'arte qual a pochi è mostra.
La quale dai suoi savi si dimostra
In Demarcana et in Bresid segrete.
Divisa con maniere ornate e liete,
E poco è intesa nell'etade nostra.
Dunque Cardan, poi ch'anco in questa parte
L'ingenio tuo fiorisce alto e adorno;
Veggio al part de gl'antichi immortalarte.
E tu Milan di lui nido e soggiorno,
Gioisci del splendor ch'a te comparte,
E riempisce ormai ogni dintorno."

Al Medesimo
"Di Girolamo Cardan medico esperto
Giunsi nel studio, al qual eran cento asse
Intorno; et egli parea studiasse
I libri, sopra quelle ognun aperto.
Et in habito mischi in un deserto
Parvemi che volando se n'andasse
Et ch'indir al suo capriccio egli inviasse
Il grande ingegno a lui dal ciel offerto.
Però prezzando l'opre di Tetelle
Di Abel, Giosef, Gieber, Bachon, Thebitte,
Zoroastro, Alchindo, Bocco et Astafone,
D'Ermete, Evande, Almadali, Zabelle,
Hiparco, Tolomeo, Aron, Beritte,
Nazabarub, Chiramide; et Trifone."

48 Lomazzo, *Rime*, p. 538:
"Doppo ciò non passaron molti giorni;
Che per grave accidente gli occhi miei
Chiuse; e perdei l'amata e cara luce
Che mi fece restar fuor di me stesso;
Si come hevea predetto il gran Cardano,
Medico et Mathematico pregiato.
Il qual ritrassi con sua sfera et libri."

the people/Since the movements of the stars are worth nothing anymore if even the numbers change and our hope is completely gone."[49]

In this rather obscure poem, the author shows an interest in the fate of astrology and astrologers, underlining the risk for astrologers to be misunderstood by people, because as he states, "When even numbers change, our hope is completely gone", a statement that testifies to Lomazzo's faith in mathematics and astrology, and the need for astrologers to convey their messages and predictions in an understandable way. As we have seen, Lomazzo indicates the exact time and place of his birth at the beginning of his autobiography, a clear indication that he knew and conveyed this information in the past to have his horoscope made. The artist does not seem to question the reliability of the observation of the stars, knowledge he probably derives from his personal acquaintance with Girolamo Vicenza, since the latter was elected "official astrologer" of the Academy of Blenio under the name of *Compà Ramozz.* He probably made the author's horoscope as well. The knowledge of planets and elements was crucial in order for the author to build his system of colors and the importance of numbers (especially of the number seven) can be seen throughout his literary production.[50]

On the more "social level", we are beginning to understand how Lomazzo often expresses in his writings direct opinions on subjects that fall beyond the scope of his artistic activity. A similar example is his position towards alchemy and alchemists, whom he often criticizes (both implicitly and explicitly) in more than one poem of the *Rime.* In some particularly relevant passages that we can find among the verses of his poems, the author explicitly criticizes alchemists, stating in passing that: "They are ignorants those who believe they can fix Mercury", or "Other fools such as alchemists, cheaters and fortune tellers", or again: "The false alchemy which misguides those who follow it."[51]

49 Lomazzo, *Rime*, 1587, p. 160:
"Furno tutti gl'Astrologhi sensati
A consiglio tra lor raccolti insieme.
Dicendo al fin il mal ch'ogn hor si preme
È l'esser de le genti abbandonati,
Poche nulla più vaglion gl'osservati
Moti del ciel da noi onde si teme
Ch'in tutto spenta sia la nostra speme
Essendosi anco i numeri variati."

50 See chapter 4: "Lomazzo's System of Colors".

51 Lomazzo, *Rime*, respectively pp. 123, 199 and 367:
"Sonvi ignoranti ancor che nell'Archimia
Si credono affissar il gran Mercurio

As in many other parts of his poetic writings, Lomazzo's opinion on alchemists is stated very clearly and directly, as if it was a result of direct confrontation and of dialogues with fellow academicians. We could find a sort of "visual compedium" of these poems in a print by a colleague and fellow member of Lomazzo's academy, Ambrogio Brambilla, who was elected chief of the academy of Blenio eight years before our artist. In his *Arboro della Pazzia* ("the tree of folly"), he puts all the men he considers to be fools under the tree, and among those, alchemists and astrologers side by side (fig. 1.2 and fig. 1.3). This print was probably inspired by the fresco by Meda and Arcimboldo in Monza, *L'Albero della Vita* ("the tree of life", fig. 1.4), although the subject depicting a wide range of "fools" is part of those *genre* of sixteenth-century prints representing the follies of human life, usually defined as 'popular prints', though, as Michael Bury already noted, this classification is in many ways problematic.[52] Certainly, discussions on topics such as the reliability of the stars and the usefulness of alchemy were circulating in the city at the time. As a result, Lomazzo's eclectic mind, interested in everything that was happening in the literary environment of his city, decided that these discussions were worth considering and that it was profitable to include his opinion on them in his writings. The interest he shows in subjects other than those related to his artistic activity, the knowledge of which he had aimed to improve since his early years, emerges clearly in the author's own voice. To what extent this knowledge was circumstantial is difficult to pin down with precision; however, what can be inferred is that he derives the notions on subjects which are relevant to his activity as a writer on art not only from reading but also, as I have hinted at in this chapter, from men of letter and science, as well as from fellow artists with whom he was personally acquainted, and who played a central role in his education and literary achievements.

Senza mirar come egli è instabil sempre."
"Saltò in campo una schiera d'altri allocchi
Come Alchimisti, Barri e Chiromanti,
et altri di mal far maestri e cuochi
Ma inanzi a questi et altri assai forfanti."
"La falsa Alchimia, che i suoi savi abbaglia
De la qual tanta stima già ne feci
Fu con gloria raccolta da tredieci
Uomini, che stentar poi nella paglia."

52 See Michael Bury, *The Print in Italy 1550–1620* (London: The British Museum Press, 2001), pp. 150–152. The astrologer depicted in Brambilla's print on the left refers to the episode of the ancient philosopher and astrologer Thales of Miletus, who was so intent upon watching the stars that he failed to watch where he was walking and fell into a well (Plato, *Theaetetus*, 174A).

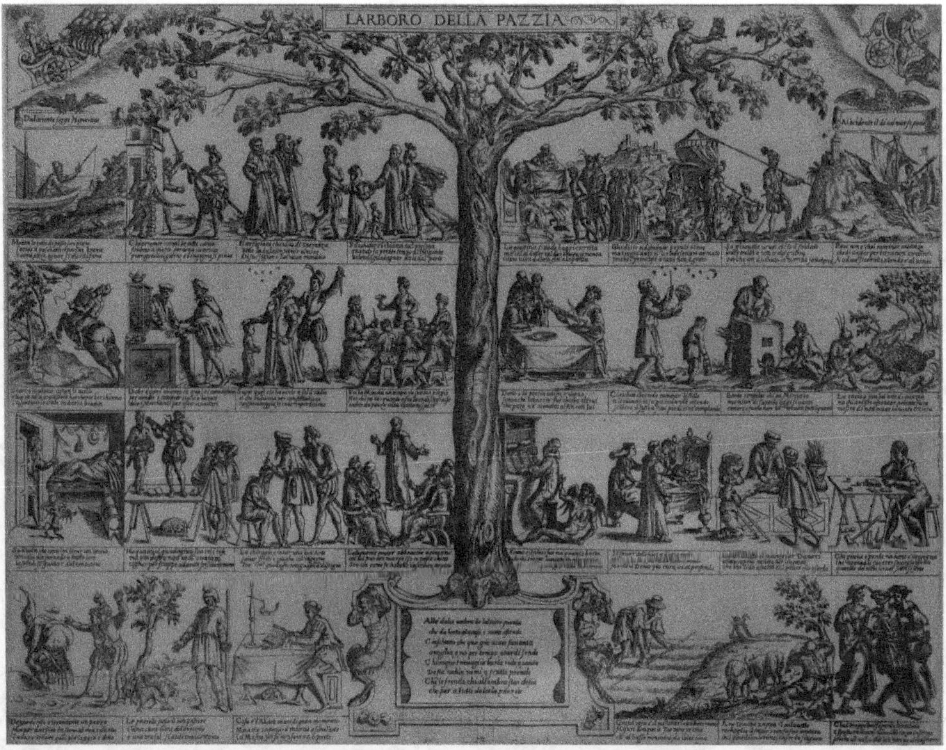

FIGURE 1.2 *Ambrogio Brambilla,* The tree of madness, *c. 1570, London, British Museum*

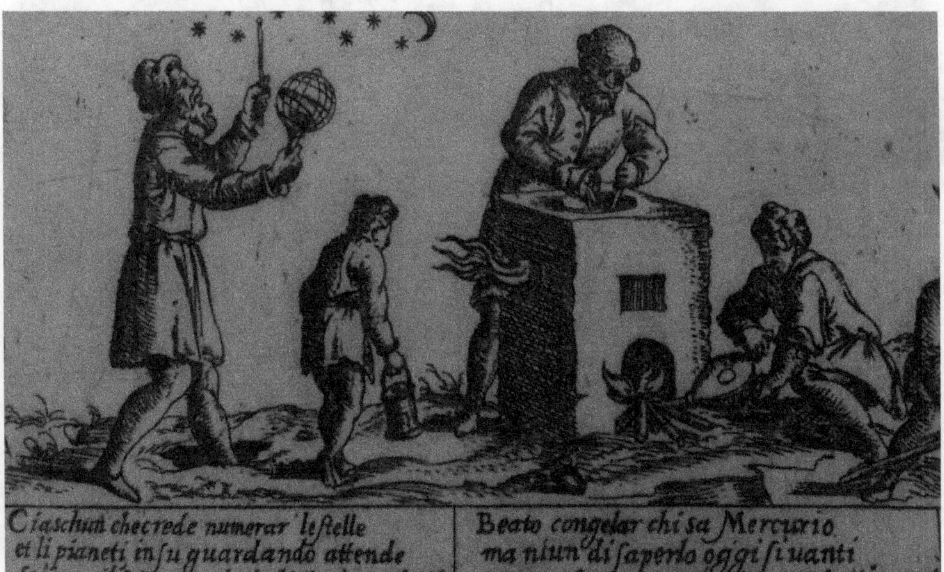

FIGURE 1.3 *Ambrogio Brambilla,* The tree of madness, *detail, c. 1570, London, British Museum*

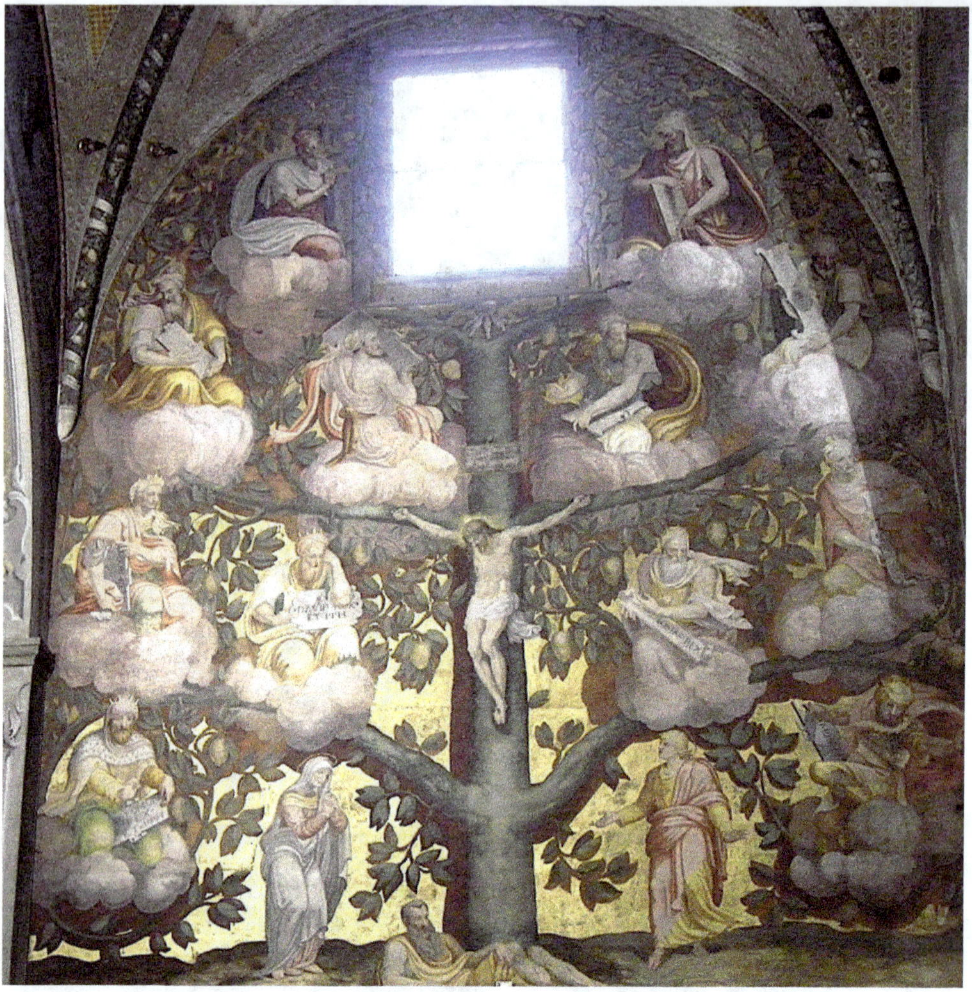

FIGURE 1.4 *Giuseppe Meda and Giuseppe Arcimboldo,* The tree of life, *c. 1550, Monza, cathedral*

CHAPTER 2

Spaces and Institutions

Milan and the Arts: Spaces of Interaction, Spaces of Production

Sixteenth-century Milan holds a fairly unique place among Italian Renaissance cities, both because of its geographical and political positions. Unlike other Northern independent cities like Venice or cities with an established court like Mantua, Milan, after being ruled for many years by the Sforza court, suddenly found itself under the Spanish domination, which started only ten years after the outbreak of the plague in 1524 and lasted for more than a century and a half.

Different scholars have brilliantly shown how the city became one of the main economic and political centers in Northern Italy during the fifteenth century. Recent works by Evelyn Welch and Monica Azzolini took into account some crucial aspects of the city's social and cultural environments, in order to understand the dynamics of power in the city under the Sforza domination. These studies focus on specific aspects of Milanese life such as the use astrology in the courts, the practice of shopping in the city, and the creation of the Sforza's image of authority.[1] Even more recently, Stefano D'Amico undertook a global analysis of Milan under the Spanish domination, enquiring on the demographic, social, economical and political aspects during the sixteenth and seventeenth centuries, providing, among other information, detailed descriptions and statistics of the socio-occupational groups active in the city from around 1576 until 1700.[2]

However, it could be generally agreed that Milan is still an understudied case compared to other cities of Renaissance Italy, especially with regard to the artistic production of the *botteghe* during the second half of the sixteenth century, the period in which Lomazzo and the artists related to him lived and

1 Evelyn S. Welch, *Art and Authority in Renaissance Milan* (Yale University Press: New Haven, 1995), and Evelyn S. Welch, *Shopping in the Renaissance* (Yale University Press: New Haven and London, 2005); Monica Azzolini, *The Duke and The Stars: Astrology and Politics in Renaissance Milan* (Cambridge: Harvard University Press, 2012). On Renaissance Lombardy (from 1454 to 1535) see also the recent study of Giovanni Romano, *Rinascimento in Lombardia: Foppa, Zenale, Leonardo e Bramantino* (Milan: Feltrinelli, 2011).

2 Stefano D'Amico, *Spanish Milan: a City Within the Empire, 1535–1706* (New York: Palgrave MacMillan, 2012).

 | DOI 10.1163/9789004330269_004

worked (from approximately 1540 to 1590).[3] In the past, different art historians argued that the artistic scene of the city supported a flourishing decorative arts industry while the figurative arts (especially painting) stagnated.[4] Undoubtedly, Milan was renowned for the excellence of its many craftsmen and artisans, who received commissions from all around Europe. The *botteghe* of crystal workers such as the Saracchi and the Miseroni families, engravers such as Annibale Fontana, or embroiderers such as Scipione Delfinone, produced luxury objects that were designed for the most important Italian and European courts, including, to mention only the most illustrious example, the court of Rudolph II in Prague (see fig. 2.1). Practitioner and writer Biringuccio describes a sixteenth-century Milanese brass-making workshop with these words: "So that whoever entered that shop and saw the activity of so many persons would, I think, believe as I did that he had entered an *Inferno*, or, on the contrary, a Paradise, where there was a mirror in which sparkled all the beauty of genius and power of art."[5]

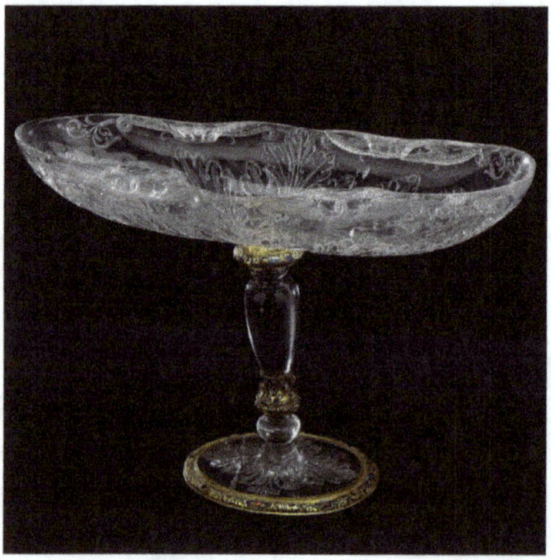

FIGURE 2.1
Miseroni or Saracchi's workshop, Tazza, *c. 1550–1580, London, British Museum*

3 As D'Amico puts it: "Despite its economic, politic, and religious importance, Milan remains the least studied of the major early modern European cities. If the scarcity of studies is particularly stunning within the Anglo American historiography, even the work of Italian scholars has come to the forefront only recently when researchers recognized the importance of the city." D'Amico, *Spanish Milan*, introduction: "A Forgotten City".

4 See the studies of Berra, Porzio and Bora, among others.

5 Andrea Bernardoni, "Artisanal Processes and Epistemological debate in the works of Leonardo da Vinci and Vannoccio Biringuccio," in Sven Dupré (ed. by), *Laboratories of Art* (Cham and New York: Springer, 2014), p. 57.

The idea that the quality and the excellence of these expensive products outshined the other arts (painting especially) is tantalizing. But in fact, upon closer inspection, it becomes clear that the artistic landscape was much more varied and complex than that. Without a doubt, in the second half of the century the city's community of painters was still under the strong influence of Leonardo's example (whose work in Milan was shown in 2011 in the excellent exhibition "Leonardo da Vinci, Painter at the Court of Milan", held in 2011–2012 at the National Gallery in London).[6] Nonetheless, if it is true that Milan witnessed a lack of innovation in painting techniques, which may have fostered as a consequence Lomazzo's theoretical discourse on the importance of painting and on the education of the artist, the argument that Milanese artists active in this period were slighted by the local community of men of letters and passed over for public commissions in favor of foreign artists such as Bernardino Campi from Cremona, is indeed mitigated when one looks for instance at Lomazzo and Annibale Fontana's cooperation for the church of Santa Maria presso San Celso, Aurelio Luini's involvement in the frescoes of the Ducal Palace commissioned by the Spanish governor Antonio de Guzman, or the numerous *pale d'altare* painted by Paolo Camillo Landriani for the convent of Sant'Eustorgio. These are only few examples that testify to the active engagement of these local artists in the city. Many of them knew each other personally and they constituted the artistic community of Milan, sometimes cooperating, often in competition with each other, and certainly interacting in environments besides the workshop's walls. As Mauro Pavesi stated and as we have seen in the previous chapter, Lomazzo himself, though frequently scorned by later critics for not being very talented as a painter, had the esteem of his contemporaries who acknowledged the quality of his pictorial work in different documents, in which the artist is called "very well-known among the most noble painters of the city."[7] It is worth mentioning that the painter in sixteenth-century Milan who had more fortune among later critics is notably Arcimboldo. His "serious jokes" and their origins have been recently re-evaluated by Thomas Da Costa Kaufmann, who contextualized the origins of his still life *invenzioni* in between the painter's pictorial fantasy and his interest for imitative naturalism.[8] Lomazzo, praising the work of this artist in more than

6 Larry Keith and Luke Syson (eds.), *Leonardo da Vinci: Painter at the Court of Milan* (New Haven and London: Yale University Press, 2011).

7 Pavesi, *Giovanni Paolo Lomazzo Pittore Milanese*, p. 5: "Inter primos civitatis pictores notissimum."

8 Thomas DaCosta Kaufmann, *Arcimboldo: Visual Jokes, Natural History and Still Life Painting* (Chicago: University of Chicago Press, 2010).

one of his writings, practically explains the technique that Arcimboldo used to form his “composite heads” in his book *Della pratica della pittura*.[9] Although it is known that Arcimboldo moved to Vienna in 1562, his link to the city of Milan remained strong and his influence is recognizable in different works of the artists belonging directly to Lomazzo’s circle, such as Ambrogio Brambilla (fig. 2.2).[10] On this topic, the exhibition *Arcimboldo* (2011) shed new light not only on the painter’s career, but also on the context of artistic exchange in Milan and beyond during the mid-sixteenth century. It presented, for example, new fundamental documents on the *Scuola di San Luca*, the corporation of Milanese painters whose gatherings at the time were mainly aimed at preventing foreign painters from having access to the city.[11]

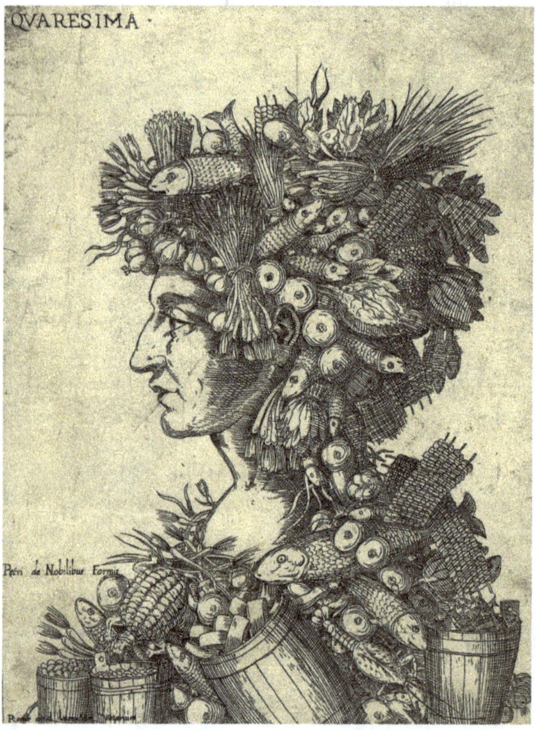

FIGURE 2.2
Ambrogio Brambilla, Allegory of the lent, *c. 1575–1585, London, British Museum*

9 Ciardi, *Scritti*, II, p. 305.

10 On the different hypotheses concerning Arcimboldo’s staying in Milan, see Charles Hope, “Sight Gags,” *The New York Review of Books*, 1987, pp. 42–44 and Thomas DaCosta Kaufmann, *Arcimboldo*, p. 243, n80.

11 See Sylvia Ferino-Pagden (ed. by), *Arcimboldo, Artista Milanese tra Leonardo e Caravaggio* (Milan: Skira, 2011), especially the essay by Robert S. Miller, “Arcimboldo e il Contesto Milanese: la Scuola di San Luca nel 1548–49 e gli Esordi del Pittore fino al 1562,” pp. 85–99.

Despite the importance of this relatively recent scholarship, the relationships and connections between Milanese artists in the second half of the sixteenth century (as well as the connections between artists and Milanese men of letters) are still, for the most part, objects of speculation. The same goes for the discussions that were taking place on the arts and on the artist's education promoted by learned painters such as Lomazzo, years before the foundation in the seventeenth century of the *Accademia delle Arti del Disegno* by Federico Borromeo (1619). In order to understand the artistic climate of the city and to get a clearer sense of the dynamics and relationships that linked the communities of artists and artisans, in this chapter I present some institutions established by artists for artists, and I examine the nature of two spaces in which they interacted: the *Scuola di San Luca* and the *Accademia de la Val di Blenio.*

Markets for the Arts: The *Scuola di San Luca*

The city of Milan in the sixteenth century was divided into a number of *quartieri* in which different arts and crafts were exercised. As was usual for Renaissance cities, every trade had its specific place or street. Paolo Morigia, in the chapter "On the variety of arts in Milan", from his *Historia dell'Antichità di Milano* (1592), states:

> It is worth seeing the union of all trades in this city. One can see all goldsmiths and jewelers gathered in two streets, all those making armors and coats in another, those who make swords and daggers and harquebusiers in another; the same can be said about the union of all other trades.[12]

Though this division would become even more defined in the seventeenth century, one can legitimately say that already in this period the *contrade* were usually subdivided according to the trades, so that many of them (the ones in which goldsmiths, arm makers, silver workers, perfumers and so on were operating) were named after the crafts that were exercised there. The number and

12 Paolo Morigia, *Historia dell'Antichità di Milano* (Venice: Appresso i Guerra, 1592), p. 260: "Inoltre è cosa degna di vedere l'unione delle arti in questa città. Laonde si veggono tutti gli orefici, e gioielleri unitamente in due strade, tutti quei che lavorano d'armature e che fanno giacchi di maglia, sono in un'altra via; quei che fanno spade e pugnali e archibugi in un'altra; il medesimo dico della unione degli altri mestieri." Translated by D'Amico, *Spanish Milan*, p. 23.

kinds of merchants and artisans active in the city was incredibly varied. Morigia also writes that:

> There are a great number of artists in different kinds of arts and I think there are very few cities not only in Italy, but also in Europe that have more kinds of *mestieri* than Milan, and a greater number of artisans. So in the office of the *provigione* of the city eighty-two kinds of arts are registered, with their *paratichi.* And besides the noble arts, one must account for the trades that are exercised inside the houses.[13]

The *ufficio di provigione* was the administrative body that wrote and published the rules for the commerce. The several *contrade delle arti*, placed in the area of the Broletto, were home to different kinds of craftsmen, such as goldsmiths and embroiderers. Usually gathered around the parishes, the concentration of the same craft in one specific place facilitated exchanges, personal relationships, and, needless to say, it enhanced competition as well, raising in many cases the quality of the final products. The prosperity of one particular trade was usually underlined by the expansion of the *botteghe* from one parish to another. For instance, as D'Amico noted, at the end of the sixteenth century the metal industry employed more than twenty-five percent of the families in the parishes of St. Michele al Gallo, St. Mattia alla Moneta, St. Satiro and St. Maria Beltrade (the last was the parish where the Lomazzo family lived, a *quartiere* famous for the diversity of artisans who worked there).[14] At the end of the sixteenth century, Milan hosted around eighty-five registered guilds of craftsmen. Yet, as D'Amico noted, since not all the artisans present in the city had the possibility or the inclination to register as members of a guild (as the membership entailed a certain amount of benefits as well as financial obligations), we can fairly suppose that that the number easily exceeded one hundred.[15]

The diversity of sellers and products that were sold in the city is well summarized in the print by the Milanese Ambrogio Brambilla entitled "A Portrait of those who go around selling and working in Rome", published in Rome in

13 Morigia, *Historia*, p. 259: "Oltre che ci sono grandissimo numero d'artisti in diversi sorti d'arti, e credo che ci siano pochissime città non solo in Italia ma anco in tutta Europa, c'habbino più sorte di mestieri di quei c'habbia Milano, ne così gran numero di artigiani. Laonde nell'officio della provvigione della città ci sono scritti ottanta due sorti d'arti, con suoi paratichi; oltre all'arte nobili, e mercanti e che s'esercitano dentro delle case."

14 D'Amico, *Spanish Milan*, p. 23.

15 D'Amico, *Spanish Milan*, p. 46.

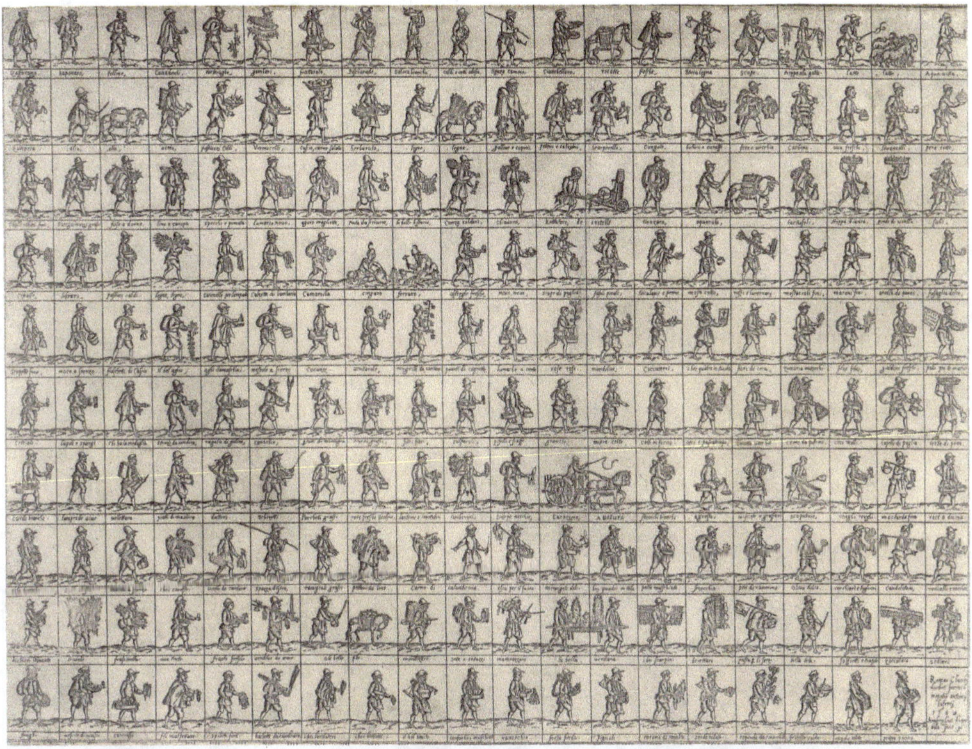

FIGURE 2.3 *Ambrogio Brambilla,* Portrait of those who go around selling and working in Rome, *1582, London, British Museum*

1582 (fig. 2.3).[16] Although the print's title refers to the city of Rome (where Brambilla, friend, colleague and fellow academician of Lomazzo, moved towards the end of the century), one cannot but notice that in the print vendors are selling products typical of Northern Italy (such as the *mostarda* and the *rane*), an indication that Brambilla actually wanted to depict the Milanese merchants and vendors whom he recalled from his hometown.

Among these vendors, we see that two are selling respectively "beautiful pictures on canvas" and "beautiful pictures on boards" (fig. 2.4). It is very possible that these two vendors refer to the selling of devotional images (as a devotional image seems to be the one that the vendor on the right holds in his hand), and of copies and drawings of famous paintings. The market and the exchange of these copies was flourishing in the city at the time, and it is in this sphere that we should interpret a contract between Lomazzo and the officer Giulio Claro

16 See Welch, *Shopping in the Renaissance*, p. 51, and Bury, *Italian Prints*, p. 166.

dated 1569.[17] In this obscure document, the jurist Claro, who was appointed by Philip II *reggente* -that is counselor for the Italian affairs in Madrid, declares that he will give Lomazzo eighty drawings by the hand of famous painters, five paintings by Titian, Michelangelo, Luino, Bramantino and Soiaro, and many other things that are his *studiolo*, valued 150 *scudi*. In exchange for these, Lomazzo will give Claro fourteen paintings by his own hand. The total price of Lomazzo's paintings (a Christ inspired by the *Cristo nel Sepolcro* by Bramantino valued 20 *scudi*, a Christ on the Cross valued 10 *scudi*, and finally, as the contract states, "Twelve figures of the barbarian kings depicted in the style and form in which Wolfgang Lazius describes them in his *De Migrationibus Gentium*", valued 10 *scudi* each), constitutes the exact price (150 *scudi*) of the paintings and drawings that Claro offered, which is far too low a price to consider these works of art, allegedly made by the famous artists quoted in the contract, as originals.[18] This contract gives us a hint on the kinds of deals painters like Lomazzo were accustomed to do and on how paintings were also used as "exchange goods".

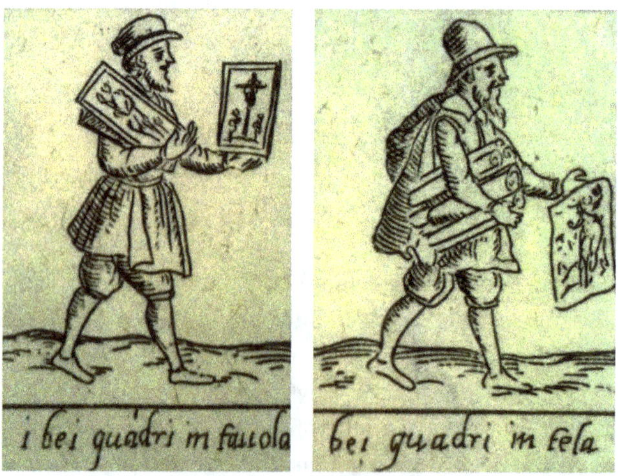

FIGURE 2.4 *Ambrogio Brambilla,* Portrait of those who go around selling and working in Rome, *detail, 1582, London, British Museum*

Painters in Milan were grouped and registered at the *Scuola di San Luca*, whose date of foundation is not known, and whose *statuti*, written in 1481,

17 Quoted in Klein, *Idea*, II, pp. 449–451. See appendix 1.

18 I thank Professor Rossana Sacchi and Professor Silvio Leydi for discussing the contract with me and helping me to make sense of it.

were unfortunately lost at the beginning of the sixteenth century. Different documents can be found on these painters' gatherings and on the members belonging to this *corporazione* in the fifteenth century.[19] As Shell already noted, *paratico*, *università* and *scuola* were usually used as synonyms in the documents, so it is often not clear to which one they are referring in each case.[20] Certainly, from 1481 painters used the term *universitas pictorum* to describe the school. In a document recently published by Robert Miller, that is a *sindacatus pictorum Mediolani* dated 9th of April 1548, we see that painters belonging to the *Scuola di San Luca* gathered together in order to re-establish the *statuti* which were lost during the years of the plague, and in this gathering we find a list of twenty-six painters, many of them not known to later critics.[21]

Among them we find Aurelio Luini and his brother Giovanni Pietro, Giovanni Angelo Lomazzo, brother of Giovanni Paolo, two other painters named Lomazzo, "Giovanni and Raffaele son of Bartolomeo", and Giovanni Crespi (the grandfather of the famous Giovanni Battista Crespi named *Il Cerano*).[22] Giovanni Paolo is not mentioned here or in the later *atti* which testify to the later meetings of the *corporazione* (respectively in 1549, 1559, 1566). This may be an indication that Lomazzo was actually not a member of the *Scuola*, although one cannot deny that the presence of his brother during these meetings may suggest the contrary, as it was not unusual to send only one person from a family of artists, and in this case Giovanni Angelo could have been the representative for the Lomazzo family (as I stated in the previous chapter, three out of four of Lomazzo's brothers were painters, the fourth embroiderer, and the father, Giovanni Antonio, probably exercised a practical activity, as he is often referred to as *magister*). Whatever the case, it is significant that Milanese painters felt the need on more than one occasion to recover their *statuti*: they were trying to find and secure their rules as well as their identity, in order to build a stronger guild that could stand up against the growing number of foreign painters who wanted to settle and work in the city.

The *Scuola di San Luca* had different ethical as well as jurisdictional rules, and overall it offered painters both protection and acknowledgment of the quality of their work. Every member of the school had to execute his commission with *bona fede et senza fraude*, that is with "good will and with no intention

19 Janice Shell, *Pittori in Bottega* (Turin: Allemandi, 1995), pp. 203–217.

20 Shell, *Pittori in Bottega*, p. 18.

21 Robert. S. Miller, "Arcimboldo e il Contesto Milanese: La Scuola di San Luca nel 1548–1549 e gli Esordi del Pittore Fino al 1562," in Ferino-Pagden, *Arcimboldo*, pp. 85–101.

22 See Miller, "Arcimboldo e il Contesto Milanese", p. 96, appendix 1.

of deceiving", especially concerning materials. The wood for the boards, for instance, had to be of good quality, and if a patron decided to fire an artist, no other artist could have substituted him without the latter's permission.[23] The *scuola* never became a *paratico*, but it preferred to always had the status of a *universitas*, that is to say that the organization had no chief, or, as they called it, abbot (*abbate*). This institution, directed and managed by one *priore* and twelve so-called *sindaci*, could not put its decisions into practice if the latters did not have the cooperation of the painters who adhered to the school.[24] Despite these limitations, the school was the prominent official administrative body that painters could refer to at the time in the city. Though the connections and the relationships between the artists who were members cannot be pinned down in detail, we can infer that the school certainly aimed at promoting local painters against foreign artists. That is especially true if one considers the competition that Milanese painters witnessed between Bernardino Campi and Giuseppe Meda in 1561, in order to establish who should have received the commission for the *ante* of the organ in the *Duomo*. This quarrel between the Milanese Meda and the foreign painter from Cremona lasted until 1565, when, after the intervention of Carlo Borromeo, Meda was finally declared the winner.[25] Quarrels like this were common at the time, and I will discuss in more detail another famous example of competition between Martino Bassi and Pellegrino Tibaldi in the chapter dedicated to perspective.

In connection with this *Scuola*, one has to look at other places where art was discussed in order to get a sense of the artistic scene during these years. The Academy of Federico Borromeo will be established only in 1619; yet, we can already see how painting in Milan was approached and referred to as "science" (following the more comprehensive Latin meaning of *scientia*, "knowledge"), not only by Lomazzo, but by all the network of people who formed in 1560 the *Accademia de la Val di Blenio*.

Spaces for Discussion: The *Accademia de la Val di Blenio*

The *Accademia de la Val di Blenio* was distinguished by the unusual language used there and by the heterogeneous character of its members, about whose activities too little is known. It was not an art academy in the proper sense of

23 See Shell, *Pittori in Bottega*, pp. 27–28.

24 See Shell, *Pittori in Bottega*, p. 29.

25 On this quarrel see also Miller, "Arcimboldo e il Contesto Milanese", pp. 87–90.

the term, yet it hosted different artists and artisans along with men of letters, astrologers, physicians and the like. The academy was founded in Milan in 1560, and named *Val di Blenio* (also spelled *Bregno*, or *Brenno*) after a valley between Switzerland and Como, part of Italy in the sixteenth century. This valley was the home of the so-called *facchini* ("porters"), men who carried food and wine to Milan. The *Accademia* was characterized by an anticlassical stance, in the sense that the participants, adopting as official language the humble dialect spoken by the porters, aimed at creating an informal place in which they could compose poetry and share their interests.[26] Its peculiar nature is self-evident from the questions a candidate had to answer in the interview to become a member, conducted by the *Gran Cancelliere*, that is, the great counsellor of the academy (*L'interogaçigliogn ch'o s'han da fá dar gran Scanscieré pos ra gneregada a colch'o vur intrò in dra Vall de Bregn*), which were all related to the job of a porter. The questions included how the straw used for packaging must be tied; of what type and how long the rope used for this should be; the characteristics of good wines, both red and white; how different implements and instruments used by the porter should be kept (including tools such as the *fusella* – "stick" and the *sciavatt* – "slippers"); how to flay a kid; and how to arrange the bags to transport a load.[27] In adapting this attitude and disguise the academicians wanted to stress their close link to the rural roots and traditions that were still solid between the city and the country at the time. In this respect, the adoption of bacchanalian rituals, as the wearing of the costumes with the attributes of Bacchus when one member was elected chief, the odes they dedicated to the pagan god, the ritual heavy drinking, are all actions inscribed into a program of reevaluation of rural traditions connected with the solar calendar and with the pagan festivals related to the grape harvest (fig. 2.5). The members called themselves *facchini*, taking this figure of the ignorant but *genuino* porter as a telling model for their anti-rhetorical attitude. Some verses by Lomazzo, dedicated to the jurist (*giurisperito*) Quinzio, clearly explain the anti-élitist stance of the *Accademia* to the otherwise unknown candidate *Nappione* ("big nose"): "You must not think, Napion, that you can become a member of the academy by your obstinacy. (…) Be careful not to be

26 This chapter is partly taken from a previous article. See Tramelli, "Artists and Knowledge in Sixteenth-Century Milan: the Case of Lomazzo's Accademia de la Val di Blenio", *Art and Knowledge - Special Issue of Fragmenta, Journal of the Royal Netherlands Institute in Rome*, 2011, pp. 121–137. Among the studies on this academy, I wish to recall the important work of Isella, *Rabisch*, pp. ix-lxii; Giulio Bora, "Da Leonardo all'Accademia de la val di Bregno: Giovanni Paolo Lomazzo, Aurelio Luini e i Disegni degli Accademici" *Raccolta Vinciana*, 1989, *XXIII*:73–101; J.B. Lynch, "Lomazzo and the Accademia della Valle di Bregno" *The Art Bulletin*, 1966, *48.2*:210–211.

27 Isella, *Rabisch*, pp. 71–7. See appendix 2.

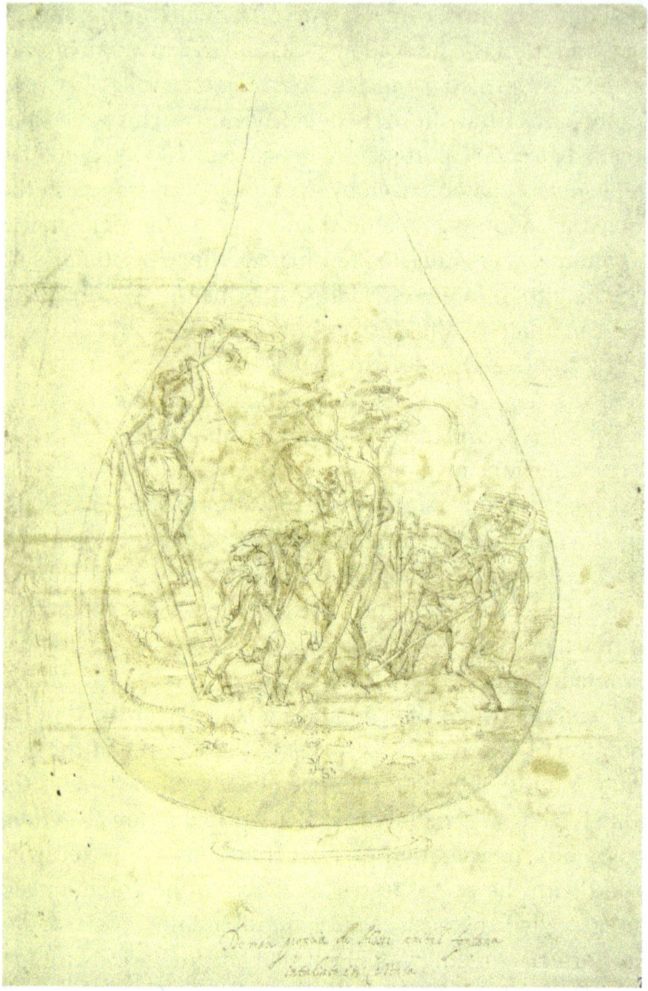

FIGURE 2.5 *Annibale Fontana,* Design for a crystal bottle, *F 245 inf, fol. 74,* ND *cat. No. 28, c. 1550, Milan, Pinacoteca Ambrosiana*

considered a fool, because, among good people, no more respect is paid to a prince than to a porter or an artisan."[28]

Moreover, the degree to which the members of the academy cultivated a humble profile is made very clear in this passage:

---

28 Isella, *Rabisch*, p. 183: "O 'n zogna miglia or mè Napiogn da begm/Pensà da intrò in sta Vall con stinaçiogn/(…) Persciò t'avis che tu te guarda begn/Da n'ess tratat anch tì par ogn minciogn,/Perchè o'n 's pòrta respett, cogl' gliust in magn,/Più ar Prinçep ch'ar fachign o a l'artesagn."

> If someone thinks that the members' profession and fame, because they changed their magnificent names into low and humble ones on entering the Valley, are as vile and fachinesque as their names, well, they evidently do not know what they are talking about. This change has been made because under the sign of humility they could show the greatness of their mind to those who think them only capable of heavy drinking and noisy chatting.[29]

The *compà* usually gathered together in the name of Bacchus in different *osterie* in Milan, drinking wine from a ritual *galeone*. The documentary evidence supports the likelihood that discussions on the education and behavior of painters took place during these gatherings, as Lomazzo's opinions on the state of the arts and on the ignorance of artists are often written in a very immediate way, as if they were the direct result of earlier confrontations. Other academies in Italy were established with similar purposes and characteristics. Some of them were said to be founded by Bacchus or Priapus, such as the Vignaiuoli in Rome or, nearer Milan, the Ortolani in Piacenza.[30]

Nonetheless, this academy can be considered one of the few examples of literary academies that had artists and artisans as leaders and counselors. It is a rarity in the context of the Italian literary academies of the sixteenth century, as these were generally reluctant to accept artists in leading roles, especially in a city like Milan, which hosted flourishing academies such as the *Accademia dei Trasformati* (1548) or the prestigious *Accademia dei Fenici* (1550), under the guidance of illustrious men of letters.[31] In the unconventional environment of the *Val di Blenio* artists and artisans were well received and had the opportunity for wide-ranging discussion with local men of letters and professionals. After the already mentioned painter and engraver Ambrogio Brambilla, who was elected first chief of the academy in 1560, the members subsequently

29 Isella, *Rabisch*, pp. 55–56: "Or gh'è dra scient ch'o 's pensa foss, per avè mudad i su nòm magnifich per intrà in dra vall de Bregn in nòm bass e da fachign, che ra soa profisiglion e ra soa influenza siglia vil e fachinesca com' and or nòm. Ma costor no sagn quòl che se scianscian perchè cosr o r'è stà facc accigliochè sòtt er segn dra umiltà pòssegn demostrà ar mond ra grandezza dor sò insciegn a confusigliogn de cogl' ch'o pensan ch'o siglien bogn se non da fà gnerefad e bagliad sempigl."

30 For the *Vignaiuoli*, see Michele Maylender, *Storia delle Accademie d'Italia*, 5 vols. (Bologna: Cappelli, 1998), vol. 5, pp. 466–467, and Lynch, "Lomazzo and the Academia della Valle di Bregno", p. 210. For the *Ortolani*, see Maylender, vol. 4, pp. 146–149.

31 For the relationship between artists and academies, see D.S. Chambers and François Quiviger (eds.), *Italian Academies of the Sixteenth Century* (London: Warburg Institute, 1995), especially pp. 105–112.

chose Lomazzo as their guide, as an attestation of the high opinion in which he was held. When the painter was elected *abate* of the academy in 1568 he wore the attributes of Bacchus, as was custom on such occasions and as his famous self-portrait now in the Brera Gallery shows (fig. 2.6).

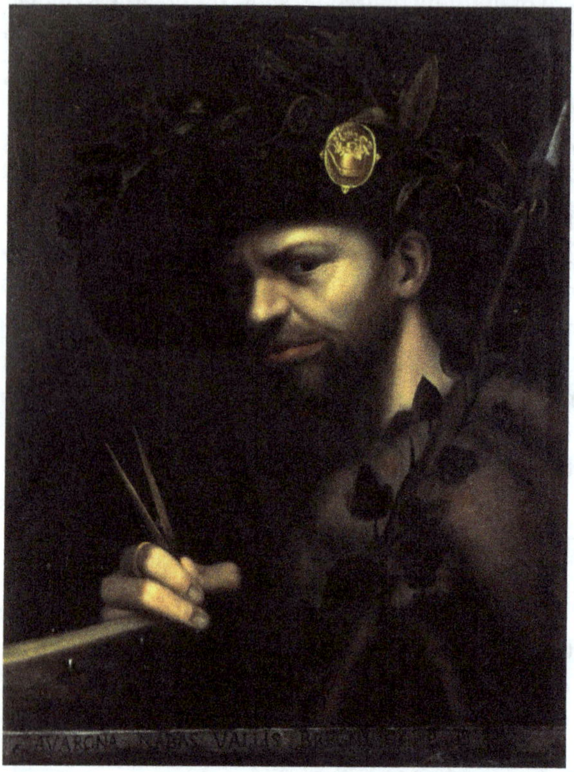

FIGURE 2.6
Giovanni Paolo Lomazzo, Self-Portrait as abbot of the val di Blenio, *c. 1568, Milan, Brera*

The anti-Apollonian god, protector of the *furor poeticus*, was chosen as an inspiration: they considered him the founder of the institution and the first inhabitant of the valley.

In his autobiography, Lomazzo talks only briefly about his leading role:

> And at that time the academy of *Blenio* was founded, and I was made Prince; there everyone spoke a rough language, and I wrote some strange poems (*Capricci*), which I may publish shortly.[32]

32 Lomazzo, *Rime*, p. 530: "E all'hor fu eretta ancor l'alta academia/Di Bregno; et io di lei fui fatto Prence/Dove parlava ognun in lingua rozza/Et io vi feci gia strani caprizzi,/Che forsi in breve si daranno fora."

The brevity of this account is in character; Lomazzo, as we have seen in the previous chapter, is also very vague about his literary education and about his apprenticeship as a painter.[33]

Lomazzo and the *Compà*

The academy attracted members from very different social backgrounds: painters, artisans, nobles, writers, physicians and the like. While precise information on the members and their place in Milanese society is scarce, the researches of Dante Isella provide valuable details.[34]

Among the artists are listed Ambrogio Brambilla (painter and engraver), Annibale Fontana (medalist, engraver and goldsmith), Scipione Delfinone (embroiderer), Ottavio Semino (painter), Girolamo Maderno (painter), Paolo Camillo Landriani, called il Duchino (painter), and Aurelio Luini (painter, son of the more famous Bernardino). Fontana and Delfinone were also among the eleven counselors (*consiglieri sapienti*). The personal relationships between these artists and Lomazzo certainly go beyond the participation in the academy: as we have mentioned in the previous chapter Fontana and Lomazzo both worked for the *fabbrica* of Santa Maria presso San Celso (though at different times), Delfinone took the younger brother of Lomazzo, Girolamo, in his workshop, and Brambilla, Luini, Maderno, along with Fontana himself, are all mentioned in Lomazzo's *Libro de' Sogni:* since the latter is dated 1563, it is probable that the painter knew all of them even before becoming *abate* of the academy (1568). Concerning the learned men, apart from the noble patron Pirro Visconti Borromeo, to whom Lomazzo's *Rabisch* is dedicated, the list includes writers, lawyers and doctors, such as Pietro Cantone, Lodovico Gandini, the Spanish Cosimo di Aldana, the Milanese Bernardo Rainoldi, Bernardino Baldini, Sigismondo Foliani and Giuliano Goselini. Other members were the musician Giuseppe Caimo, the *giureconsulto* Francesco Giussano, the engineer and architect Giacomo Soldati and the astrologer Girolamo Vicenza.

From the contents of the poems would appear that the members of the academy were to some extent critical of the society of their time, and that their *riunioni* were gatherings of relatively learned local artists and professionals who wanted to share their passion for poetry and their opinions on various topics in a congenial and informal environment.

---

33 Lomazzo, *Rime*, p. 529: "A mastro me n'andai fino a dieci anni/Dove apprendei a legger et contare/E'l maneggiar de i libri, et poi disegno."

34 Isella, *Rabisch*, pp. 329–371.

Adopting the ironic and anti-academic figure of the porter and changing their names into "humble ones", the members were free to express their passion for poetry and their opinions on the state of the arts in a relatively informal association. It probably began as a group of friends who felt the need to have an institutionalized place to pass time enjoyably outside the workshop, inviting men of letters to join them, drinking wine and presumably preparing the *apparati* for the festivals in Milan. They wrote down detailed and peculiar rules of membership and behavior, as was customary in conventional Italian literary academies. However, these rules did not explicitly exclude any person on the basis of class or social status: the criteria for the admittance of a member were established according to his literary education, that is to say, to his writing skills. The position of this academy in an area between the cultivated and the burlesque is not unique: the *Ortolani* and the *Vignaiuoli* mentioned above are only two examples of a tradition in which burlesque academies such as the *Beoni*, the *Mantellaccio*, the *Spensierati* or the *Granelleschi* can also be included.[35] Outside of Italy, the tradition of burlesque gatherings (not yet called academies) goes back to the medieval period, especially in connection with the troubadours in France and later the *compagnies joyeuses* in the Southern Netherlands, which were both related to the context of theatre.[36] It is in this context that we should interpret other prints by Brambilla, such as *Il Paese della Cuccagna* (see fig.11) or the *Scenes with Zanni* (fig. 2.7). These prints display burlesque themes closely connected with the Italian tradition of the *Commedia dell'Arte*; the first one displays the idyllic fictitious country of *Cuccagna*, where "the more you sleep the more you earn", full of rivers made of wine and mountains made of cheese, a subject which is also a reminder of the game related to the *albero della cuccagna*, a traditional pole with food and gifts that was displayed during the festivals of the city; the second one presents separate actions of the typical Italian characters (in this case Zanni the servant) that were played by the actors of the various travelling theatre companies that enlivened Northern Italy at the time.

The *Rabisch*'s poetry is in line with the burlesque literary tradition flourishing in the 1540s and 1550s, mixing fictitious dreamlike episodes with praises from one member of the academy to the other. The authors used swearwords and made jokes as a mean of being critics of the society of their time. The print

35 Francesco Cherubini, *Dizionario Milanese-Italiano* (Milan: Imperial Regia Stamperia, 1840), p. 83: "Un'Accademia poco diversa da quelle dei Beoni, del Mantellaccio, degli Spensierati, dei Granelleschi, e di tante altre cosiffatte spontanee filosofali consolatrici del nostro nonnulla."

36 On the subject, see Arjan Van Dixhoorn and Susie Speakman Sutch (eds.), *The Reach of the Republic of Letters* (Leiden: Brill, 2008), especially pp. 17–33 and pp. 79–119.

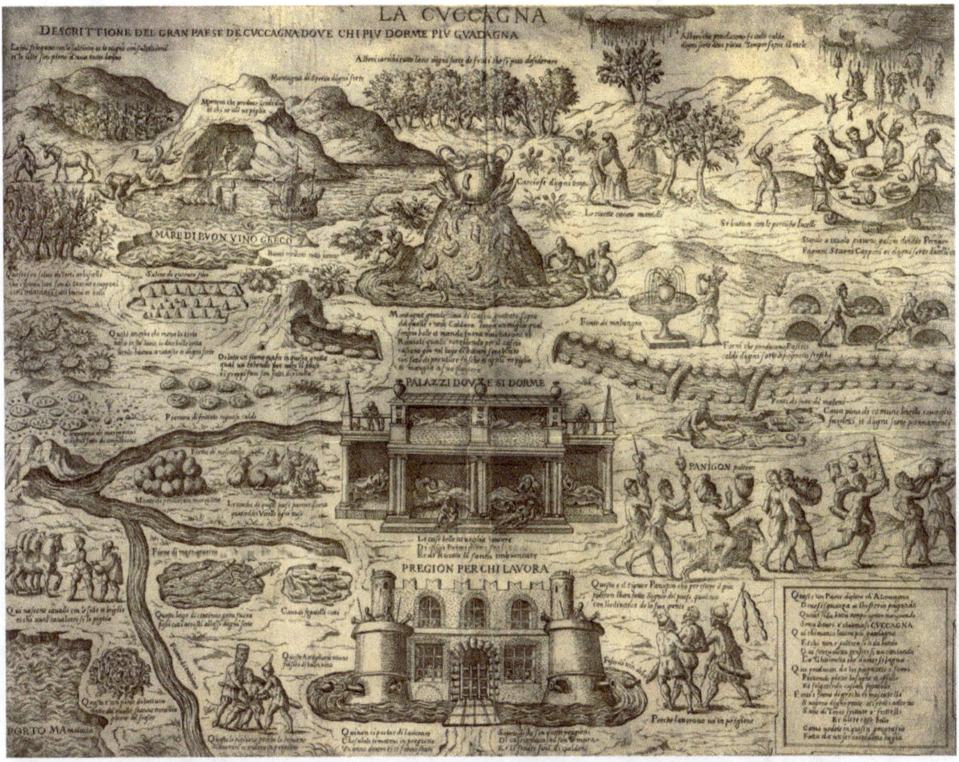

FIGURE 2.7 *Ambrogio Brambilla,* Description of the land of Cuccagna, *1575–1590, London, British Museum*

by Brambilla recalling Arcimboldo is one of the examples of this ironic attitude. Arcimboldo, as we stated before, was working for the courts in Vienna and Prague, but always in contact with the Milanese artistic and literary culture, comprising members of the academy.[37] One prominent member in particular, that is the patron Pirro Visconti Borromeo, was fascinated by the ironic and grotesque stance of the Academy, especially during the years in which he undertook a series of renovation works for his Villa in Lainate, outside Milan. As we shall see in the next chapter, the Villa was a place where grottos, nymphs, *naturalia* and *artificialia* were not only the predominant decorations, but they also constituted, as Alessandro Morandotti notes, a sort of visual counterpart to the grotesque poems in the *Rabisch.*

37 See, among others, the already quoted recent study by DaCosta Kaufmann, *Arcimboldo*.

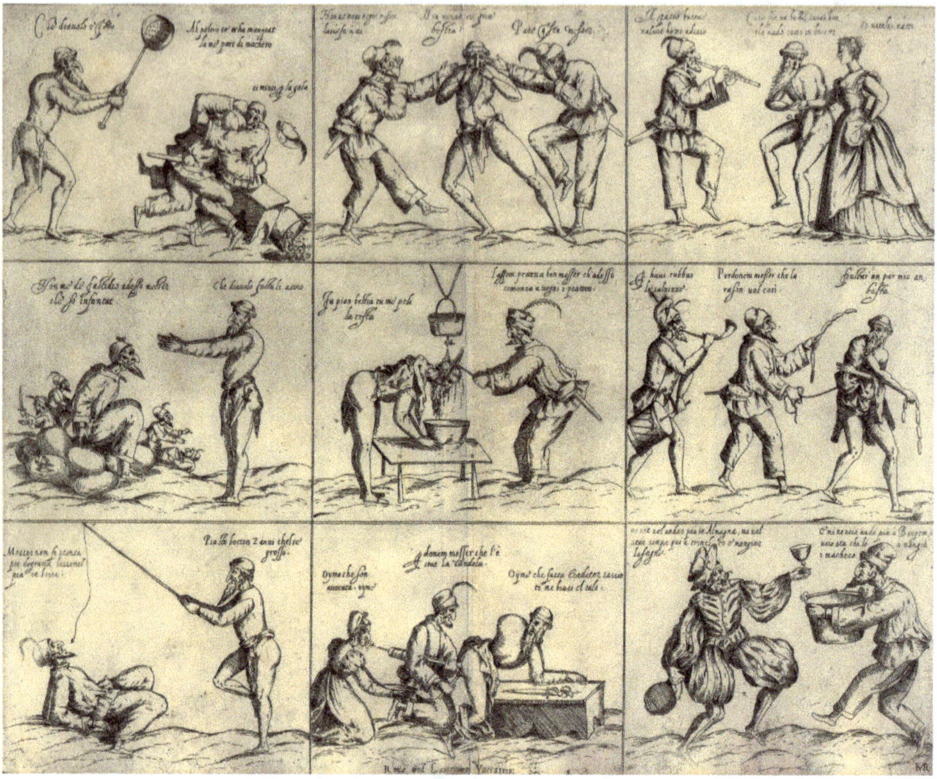

FIGURE 2.8 *Ambrogio Brambilla,* Scenes of the commedia dell'arte, *c. 1570, London, British Museum*

The Fachinesque Dialect

The *Rabisch*, a collection of poems written by members of the *Accademia*, constitutes the main source on the participants of this institution. Although the poems were probably composed earlier, the book was not published until 1589 by Paolo Gottardo Pontio.[38] The largest part of it is written in *lingua facchinesca*, that is a particular Milanese dialect, characterized by a distinct orthography and a striking rustic quality. This choice underlines the ironic, anti-academic stance adopted by the institution. The frontispiece of the book alone testifies to the freedom the academicians aspired to. As Mira Becker in

38 I agree with Isella, who suggests that the poems cannot have been composed later than 1568, when Lomazzo was elected abbot of the academy. The presence in Milan of the pious reformer Carlo Borromeo during those years may explain why the book was published quite late. See Isella, *Rabisch*, pp. xxxiii–xxxiv.

her recent article puts it: "The frontispiece of the Rabisch shows a naked male figure standing on a grass field, looking sideways, and in the Facchinesco dialect it is written 'Guarda chi or vis it me Furò ch'in Dupra (First look at my face, then look at my furor within the work). Fury is to interpret as a mixture of divine inspiration and natural talent, it releases the virtuous creativity and it constitutes a fundamental characteristic of the self-description designed for the academy."[39]

The *Bregnese* was the traditional dialect used by the *facchini*. In fact the dialect that the academicians used was not an accurate imitation of the oral language spoken by the porters: it was inspired by this dialect, but predictably polished, corrected and perfected for literary purposes.[40] At the same time it is fairly clear that the *Rabisch* was most probably meant to be read only in the context of the academy by its members: it was not addressed to a wider public, at least not outside the Milanese area. If we consider the number of publications in Milan in the sixteenth century, we find, predictably, that very few were in dialect, which was normally spoken rather than written, and, of course, was not a "language" accessible to many.[41] Although at the end of the

39 Mira Becker, "Grottesco & Suavitas: Zur Kopplung von ästhetischem Programm und institutioneller Form in zwei Mailänder Kunstakademien der Frühen Neuzeit", pp. 415–440, in Cancik-Kirschbaum, Eva and Traninger, Anita (eds.), *Wissen in Bewegung. Institution – Iteration – Transfer* (Wiesbaden: Harrassowitz Verlag, 2016), p. 421: "Das Frontispiz der Rabisch zeigt eine nackte männliche Figur, die auf einer Grasfläche steht, zur Seite schaut und zu der im facchinesco geschrieben steht: „Guarda chi or vis er me furò ch'in dupra" („Zunächst, schau hier mein Gesicht, im Werk drin sodann meinen furor"). Furor ist dabei zu verstehen als eine Mischung aus göttlicher Inspiration und naturgegebenem Talent, die virtuose Kreativität freisetzt und als grundlegendes Merkmal der Selbstbeschreibung der Akademie ausgestaltet wird."

40 Dialects are a deep-rooted part of the oral tradition of each Italian region, and they are typically unwritten. Any effort to transpose a dialect onto paper, now as then, necessarily implies corrections and modifications. The *Bregnese* especially to an Italian ear may recall a mixture of the Milanese and *Bergamasco* dialects spoken in Italy nowadays. In any case, as J.B. Lynch points out, translating the *Bregnese* involves a wide range of philological problems connected with the necessity of not losing the flavor of this language (Lynch, "Lomazzo and the Academia della Valle di Bregno.) For a discussion on the *facchinesco* dialect, see Ferdinando Cesare Farra, "Annotazioni Relative al Dialetto Usato dalla Cinquecentesca Accademia della Val del Blenio," *Rendiconti del (R.) Istituto Lombardo di Scienze e Lettere,* 1951, LXXXIV:289–307, and Jean Buchmann, *Il Dialetto di Blenio* (Paris: E. Champion, 1924). For an account on the *Blenio* valley, see Raffaello Laorca, *Le Tre valli Stregate: Documenti per la Storia delle Streghe nei Baliaggi Svizzeri di Riviera, Blenio e Leventina: 1641–1676* (Locarno: A. Dadò, 1992).

41 See Ennio Sandal, *Editori e Tipografi a Milano nel Cinquecento,* 3 vols., (Baden-Baden: Korner, 1977–1978), vol. 2, pp. 61–95. Gottardo da Ponte (or da Pontio), editor of the

book we find a table which explains the more obscure terms of this language (including an explanation of the *tavola* itself), it is still unlikely that such a book could actually have been read by people from outside Milan. And even in the case of those capable of understanding the language, there is not much chance that the topics themselves would have been of interest: most of the poems consist of praises and jokes among the members.[42]

Presumably the academicians were perfectly able to write in Tuscan, but they deliberately chose this Northern dialect. In what we could call his "academic guidelines", Lomazzo sought a revaluation of regional differences, as he makes clear in the poem to the *Fellow Baldign*, the *letterato* Bernardino Baldini:

> I understood, my friend *Baldign*, that some people, who think of themselves as the noblest men of Milan, want to study Tuscan and repudiate the country in which they were born. I see that they are heading to lose what they have by birthright, in order to get what they know they will never possess, no matter how many books writers have written. These people show that they are not worthy to enjoy the goodness of their country, as we all, porters of Blenio, do.[43]

This statement underlines that the academicians were particularly proud of their Milanese roots, and sought to fight against the predominance of the Tuscan language in an area that may have been characterized by a certain level of provincialism compared to other regions of Italy at the time.

At the beginning of the *Rabisch*, in the section labeled the *Origen e fondament dra vall* ("Origin and fundament of the valley") the disposition of the members is underlined with these words: "They need to avoid the company and the conversation of ignorant and vicious people, such as those who cannot

---

Rabisch, published between 1501 and 1548 sixty-eight books in Latin and forty-two in the vernacular, whereas we do not find works in dialect. Francesco Cherubini, in his *Dizionario Milanese-Italiano*, p. 83, mentioning the *Accademia de la Val di Blenio*, states that different sonnets written in the *facchinesco* dialect can be found in other books as well: "Anche in altri libri di quell'epoca si veggono sonetti o versi stampati in quel dialetto, semiprove del continuare in fiore quell'Accademia." The nature of these books is not known.

42 For the explanation of the *tavola*, see appendix 3.

43 Isella, *Rabisch*, p. 126: "Ho intes, Compà Baldign, che sciert Signó/Ch'o' se tegnen di più nobel de Miragn/Vun studiglià sora al parlà Toscagn/E renegò ra patriglia ond ign nassù./O vegh begn che costor o gl' ign perdù/A perd or sciert ch'han per natura in magn/Par avè quòl che magl'avè no sagn/Par quanc liber han magl'facc i scrició."

see the light because they are buried in the darkness. So they will stay away from this presumptuous and infected genie."[44]

All the members of the academy were literate and educated to some extent, since the statutes (*straducci*) required that participants be able to write: "He who ever wants to enter the valley must show what he can do with a pen in his hand. No person shall be admitted unless he is erudite in some field, principally in the liberal arts."[45]

Moreover, the members had to demonstrate to Bacchus that they were "true lovers of every science, capable of answering to anyone on any topic, with probable or evident arguments, notwithstanding the existence of many *scannaparole* and *squarciafogli*."[46]

We can here witness, *in nuce*, the emergence of the new figure of the lover of every art, the *dilettante*, embedded in those members of Blenio who were neither artists or artisans, but who wanted to know more about art and to learn how to talk about it. While the members disguised themselves taking new names and presumably talking to and discussing with one another in vernacular, the choice to write their poems in the fictitious dialect as well as the decision to include an explanatory table at the end of the book are indicative not only of the pride of the members for their own Northern origins, but also of the nature of the academy itself, in between the *serio* and the *comico*. The academicians certainly did not take themselves too seriously, and yet in the playfulness of their attitude towards the costumes and especially towards the language that they invented one can detect a special interest for literary *invenzioni*, that is, a will to play with words in the same fashion that they played with images, creating jokes and non-sensical rhymes, as Brambilla does in his print entitled *Concetto d'un Amante Uscito dele Pene d'Amore*, a forerunner of the contemporary puzzle-pictures (fig. 2.9).

44 Isella, *Rabisch*, pp. 54–55: "Ch'o deben con tutta la dirigençiglia schivà e fuscì ra compagniglia e conversaçigliogn de scient viçigliosa e ignorantera come quòlla ch' non pò guardà or s'ciaró per ess sotterrada in di scuritad dra nòcc, e par cost a sta scient prosontugliosa e amorbada o 'gh daran dra ara."

45 Isella, *Rabisch*, p. 67: "Ch'o 's mostra tutt col ch'o 's sa fà con ra penna in magn da col ch'o vul intrà in dra vall. Ch'o no s'ascieta onzugn s'or no n'è virtuglios in qualcòssa, e prinçipalment in degl'art liberal." For the complete list of the *statuti*, see appendix 4.

46 Isella, *Rabisch*, p. 56. The words *scannaparole*, literally "words' murderer" and *squarciafogli*, "papers' ripper," cannot be translated properly without losing the flavor of the language.

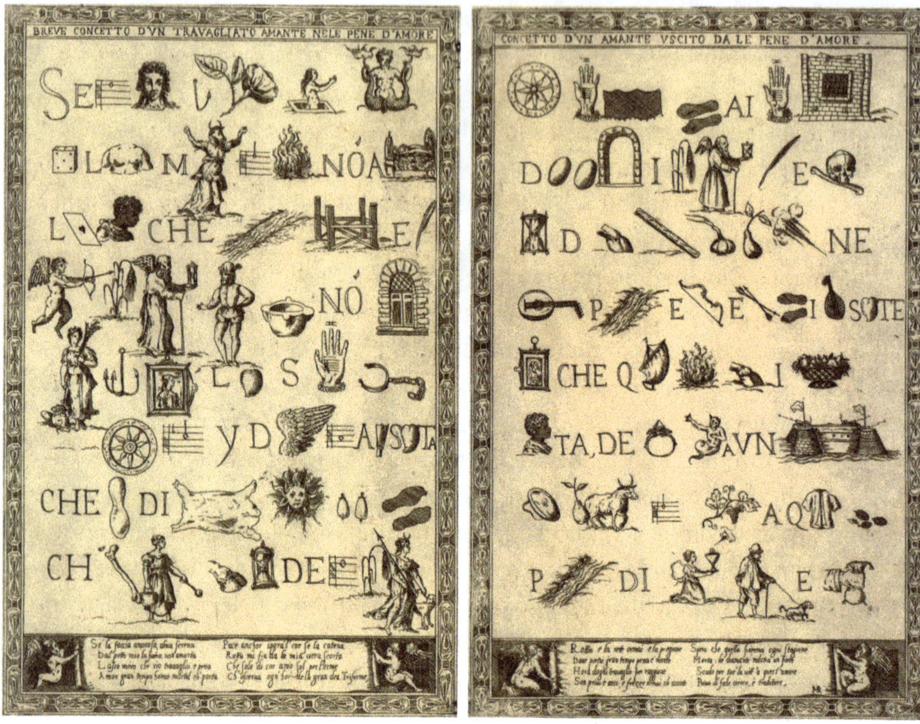

FIGURE 2.9 *Ambrogio Brambilla,* Concept of a lover, *c. 1575, London, British Museum*

The Academy in Borromeo's Time

One of the poems in the *Rabisch* dedicated to Lomazzo indicates that its author, the painter Brambilla (*Compà Borgning*), spent some time in prison because of his participation in the academy, and he warns the others to be careful "not to talk with anyone unless his character is well known."[47] Although it is possible that Brambilla was put into prison, one must acknowledge the possibility that he is only telling a fictitious story, as Lomazzo often does in his *Libro de' Sogni*, especially if one considers the ironic tone of the poem. Brambilla states: "Think, Zarvagna (while I was in prison) I went out of my mind, and some thoughts came to me to repudiate the Valley and to eat roast, but your writings comforted me."[48]

47 Isella, *Rabisch*, p. 231: "E che no faghé gnieregh con nessugn/Se prima no gl' hign begn interogad."

48 Isella, *Rabisch*, pp. 228–229.

One may think that this is nothing but a literary game between the two painters-poets, especially when one looks at all the metaphors and learned quotations that Brambilla embeds in his poem for Lomazzo: "I hoped it was that time of metamorphosis still, when Jupiter, Diana and Apollus used to feel pity and to turn men into fountains, plants and bricks."[49] That members of the academy had to suffer a certain amount of intolerance is possible, but it is also true that their gatherings were held in public (informal) places such as the *osterie*, places of entertainment and trade at the time.[50] Had the institution formally been banned by Borromeo or by the Milanese law, the gatherings could not have taken place in the city at all. For the publishing of the collection of their poems is quite a different story: in the poem dedicated to Girolamo Vicenza (the official astrologer), Lomazzo expresses the wish that after the death of Borromeo the members of the academy will be able to publish a book "full of their writings" –the *Rabisch* itself.[51] It is likely, in this case, that the approval for the publishing of such a "pagan" book would have been denied under the strict post-tridentine law, but this does not come as a surprise, as also Borromeo's attempts to banish the city's pagan rituals and traditions related to the celebration of "rural deities" are well known. The number of scholars who devoted their studies on the religious situation of Milan under the influence of the pious reformer, who was elected archdeacon of the city in 1565, is indeed great.[52] And yet, to what extent the Counter Reformation and its precepts promoted by Borromeo actually constituted an obstacle and a limitation as to what these artists could achieve in their work beyond public religious commissions is still to be defined. The argument first conveyed by Isella (1993) and then took up by Morandotti (2005), that the oppressive climate in the city under the pious reformer was suffocating the secular arts, thus creating a "hidden reaction" which the artists belonging to Lomazzo's circle should be situated in, is being, if not reconsidered, at least now taken with a grain of salt. Borromeo notably arrived in Milan in 1566, setting out to preach sobriety and humility to his community, and his attitude towards the morality of the people in the city

49 Ibidem.

50 For an account on the Milanese *osterie*, see D'Amico, *Spanish Milan*, p. 21.

51 Isella, *Rabisch*, p. 235: "E fu 'd sta scient or Lucca/Impirem on librasc di nòst scriciur."

52 Countless studies on this central figure in the field of religious studies could be quoted here; I mention only the most recent Italian studies that were written to celebrate the fourth centenary of Borromeo's canonization (1610–2010). Among them: Danilo Zardin, *Carlo Borromeo, Cultura, Santità e Governo* (Milan: Vita e Pensiero, 2010); Cesare Bonino, *La Vita e i Miracoli di San Carlo Borromeo* (Milan: Jaca Book, 2010); Dionigi Tettamanzi, *San Carlo e La Croce* (Milan: Ancora, 2010); F. Rossi di Marignano, *Carlo Borromeo, un Uomo, una Vita, un Secolo* (Milan: Mondadori, 2010).

became stricter after the plague that ravaged the city in 1576 (the second outbreak in less than half a century). His *Memoriale ai Milanesi* (1579) was notably written to remind his diocese of the examples of saints and martyrs of the Catholic Church, after he observed that the plague did not produce the effect he had hoped it would, not having helped convert the people to a more holy and religious life. The following passage characterizes his attitude towards the loose and worldly habits of the community:

> But you, Milan, are blind, so we need to speak especially to you, because of the duty we have towards you, and because you have indulged in sins and profanities more than other cities and people. Pray do remember how dangerously many people, enemies of the Cross of Christ, were conducted under the slavery of Satan, of the flesh and of the world.[53]

Borromeo's remedy against the dangers of the flesh and of earthly pleasures was the strict adoption of the Counter Reformation's precepts, including the banning of different books and public manifestations, especially those linked to previous pagan traditions, like the celebration of the harvest festivals that were taking place around the canals of the city (*navigli*) and the famous Carnival. However, it is not legitimate to infer that Borromeo was overall against the arts or against painters living in the city. Apart from the famous *Instructiones Fabricae et Suppellectiles Ecclesiasticae,* published in 1577 concerning the architecture and decorations to be adopted for churches following the decree of Trent, Borromeo also had an interest in anatomy as well as in the depiction of the human body, different anatomical treatises were dedicated to him, and he also possessed many books on this subject in his own library, such as Realdo Colombo's *De Re Anatomica* (Venice 1559), the already mentioned *De Humanis Corporis Fabrica* by Andrea Vesalius, (Basel 1543) and the *Opuscula Anatomica* (Venice 1571).[54]

53 Carlo Borromeo, *Memoriale ai Milanesi,* Milan 1579, in Nicola Raponi and Angelo Turchini (eds.), *Stampa, Libri e Letture a Milano nell'Età di San Carlo Borromeo* (Milan: Vita e Pensiero, 1982), pp. 27–28.

54 See *San Carlo e il suo Tempo. Atti del Convegno Internazionale nel 4° Centenario della Morte* (Milan: Storia e Letteratura, 1986), pp. 173–174. On the anatomical treatises owned by Carlo Borromeo, see also Agostino Saba, "La Biblioteca di San Carlo," *Fontes Ambrosiani,* 1936, *XII*:13–20. The Cardinal's interest in anatomy will be further discussed in chapter VI.

The Fate of the Academy

Concerning later sources on the Academy of Blenio, as Maylender reports, both the catalogue of academies at the end of the *Specimen Historiae Academiarum Eruditarum Italiae* (Leipzig 1725) and the catalogue of the *Conspectus Thesauri Litterarii Italiae* (Hamburg 1730) mention the *Accademia della Valle di Bregno.*[55] In his *Storia della Letteratura Italiana,* Girolamo Tiraboschi states:

> I shall not consider the academies called Fenicia, the Eliconia, and others, and the one called the Academy of the Bregno Valley, of which Gian Paolo Lomazzo was prince and in which poems in the language of that valley (called Facchinesca) were recited. Of these academies we have scant and uncertain information.[56]

Less than a century later, Francesco Cherubini in his *Dizionario Milanese-Italiano* explained, under the word *Fachin,* that:

> From 1500 onwards, at the time when Giovanni Paolo Lomazzo lived, a sort of academy was founded with the aim of combining poetry with amusement. (…) In that academy poems were written in the dialect of the Blenio valley, one of the valleys of the Italian part of Switzerland. The *Abati* were the painter Lomazzo, the "fellow Borgnin", better known as the painter Brambilla, and so on.[57]

At the moment, we do not know for a fact how long the academy lasted. If we take it for granted that Lomazzo was the predominant figure, then it is unlikely that the institution could have endured long after his death. In any case, given

55 Maylender, *Storia,* V, p. 421.

56 Girolamo Tiraboschi, *Storia della Letteratura Italiana,* 9 vols., (Florence: Molini, Landi and Co., 1805–1813), vol. 7, p. 188: "Io lascio in disparte quella de' Fenicii, la Eliconia, ed altre, e quella detta della Valle di Bregno, di cui fu Principe Gianpaolo Lomazzo, e in cui recitavansi componimenti poetici nella lingua propria di quella valle, che volgarmente dicevasi Facchinesca, delle quali abbiamo scarse e incerte notizie."

57 Cherubini, *Dizionario Milanese-Italiano,* p. 83: "Fin dal 1500 e ai giorni del celebre pittore Giovanni Paolo Lomazzo, si stabilì fra noi una specie di Accademia diretta a congiungere gli studi poetici con l'onesto spassarsi. In quella specie di Accademia si volle per ispasso poetare nel dialetto della valle di Blenio, una delle valli della svizzera italiana. (…) Abate della Valle s'intitolava il pittor Lomazzo medesimo, Compà Borgnign il pittore Brambilla, e vie va discorrendo."

that academies often survived for no longer than few years, the fact that the *Blenio* lasted almost three decades (if, as seems reasonable, we suppose that it was still going when the *Rabisch* was published) seems quite a remarkable indication of the strong connections that bound its members together. Some of them, such as Giovan Battista Visconti, Giacomo Antonio Tassani and Pietro Cantone, joined the *Accademia degli Inquieti* or *Sforzesca* in 1594, of which we have records until 1609.[58] This may well be an indication that the peculiar *Accademia de la Val di Blenio* had been disbanded by the beginning of the seventeenth century. As has been shown, the literary achievements of these academicians collected in the *Rabisch* give the best information on the members and on the nature of this institution, and although any account of the activities of the academy itself and the nature of the discussions inevitably involves a degree of speculation, the poems and the passages analysed provide valuable hints about the atmosphere of debate and the topics which interested this heterogeneous community (including Lomazzo) at that particular time in Milan.

58 Maylender, *Storia*, III, p. 303. See also Morigia, *Della Nobiltà di Milano* (Milan: Pacifico Pontio, 1595), pp. 297–299.

CHAPTER 3

Art and Grotesque

Lomazzo's Connections with the Patron of the Blenio Academy Pirro Visconti Borromeo

Among the relationships between the members of the academy, one is certainly worth investigating further—the connection between Lomazzo and the patron of Blenio, Pirro Visconti Borromeo (1560–1604).[1]

In 1589 the heirs of Pietro Tini published in Milan *L'Idea del Giardino del Mondo* ("The Idea of the Garden of the World") a book written by the physician and academician Tomaso Tomai, and dedicated to Pirro Visconti, a book "in which many of the secrets of Mother Nature are discussed."[2] They admit that the book would have reached perfection if it had mentioned the "vague and beautiful Fountain that His Excellency ordered for his most noble building of Lainate, where in addition to the most ingenious *artificio* of water one can admire the noble building with niches full of beautiful statues, and many paintings which make the place a delectable marvel."[3]

Pirro Visconti Borromeo, described by Paolo Morigia in 1592 as "one of the richest and most noble knights of the city", is an intriguing member of the Milanese nobility at the end of the sixteenth century.[4] Son of Fabio I Borromeo, of a secondary branch (*ramo cadetto*) of the Borromeo family, and of Costanza Trivulzio, he is perhaps now regarded as a minor personality in the Milanese context. At the time, however, he was a renowned patron and promoter of the arts, as well as a fervid collector of paintings and a connoisseur of the marvels of nature, which he wanted to reproduce and display in his villa just outside of Milan, in a rural village called Lainate. He was member and patron of the

1 The most complete study on Borromeo and his villa in Lainate was published by Alessandro Morandotti. See Morandotti, *Milano Profana nell'Età dei Borromeo* (Milan: Electa, 2005). A short article taken from this chapter entitled "Art and Grotesque, Art with Grotesques: The Case of Pirro Visconti Borromeo's Villa in Lainate" is in preparation for Matica Srpska Journal of Fine Arts (2016).

2 Tomaso Tomai, *Idea del Giardino del Mondo*, (Milan: Pacifico Da Ponte ad Instanza haer. Pietro Tini, 1589), preface.

3 Tomai, *Idea*, preface.

4 Morigia, *Historia*, p. 236. For a short biography of Pirro Visconti Borromeo see Isella, *Rabisch*, pp. 368–371. For a complete enquiry on this figure see Alessandro Morandotti, "Pirro Visconti Borromeo di Brebbia, Mecenate nella Milano del Tardo Cinquecento," *Archivio Storico Lombardo*, 1984, CVII:115–162.

 | DOI 10.1163/9789004330269_005

Blenio Academy, and the chief promoter of a series of renovation works for his villa (1585–1589), coordinated by the architect Martino Bassi (whom, as we shall see, Lomazzo knew personally), the sculptor Francesco Brambilla and the Parmesan painter Camillo Procaccini (1561–1629). Girolamo Borsieri describes part of the building, and especially the Nimpheo, as: "A curious and vague building, as noble as a civic building, which can compare to any other noble building, even grander, that one can see."[5]

Pirro probably decided to renovate his villa in consultancy with the members of the academy of Blenio. The result, as Alessandro Morandotti stated, was a "visual compendium" of the style and nature of the poems composed by the academicians, with decorations, arabesques, friezes and fictitious stones and wonders of nature.

As a collector of *naturalia* and *artificialia*, Borromeo was not only an admirer of the arts but also of the sciences. He indicated this by his choice to have the figure of Mercury depicted at the center of the main ceiling in the villa as protector of his home (fig. 3.1).

Procaccini, painter from Parma who also received many commissions from cardinal Federico Borromeo, decorated all the ceilings of the villa. The frescoes represent myths, grotesques, scenes from the natural and the fantastic worlds (fig. 3.2).

The most impressive building the count decided to construct was the so-called "Nimpheo" (house of the nymphs), which was placed in a central position, adorned with grotesques and statues of Fauns and Nymphs and internally enriched with different natural and artistic curiosities (fig. 3.3).

The building hosted Borromeo's collection of paintings, fossils, minerals, coins, sacred relics and archaeological findings. As Morandotti pointed out, the paintings shown in the Villa were "creating a dialogue" with the other objects of the collection, that is to say that Pirro was a thoughtful collector of both natural and artistic objects, some of which we find in the inventory dated 1604. It is plausible to infer that he was advised in his choice of items by the members of Blenio, and probably by Lomazzo himself. However, this kunst-or wunderkammer aspect was not fully recorded at the time by viewers such as the man of letters Girolamo Borsieri, who was more interested in recording the careers of the painters from Lombardy at the time. He states that: "Pirro Visconti collected some paintings, not many but so exquisite that

5 Paolo Morigia, *Il Supplimento della Nobiltà di Milano*, edited by Girolamo Borsieri (Milan: Giovanni Battista Bidelli, 1619), p. 60: "Ha vaghezza di edificio curioso, nobiltà di civile (...) Che può paragonarsi con qualsiasi altro edificio anche maggior che oggi si vegga."

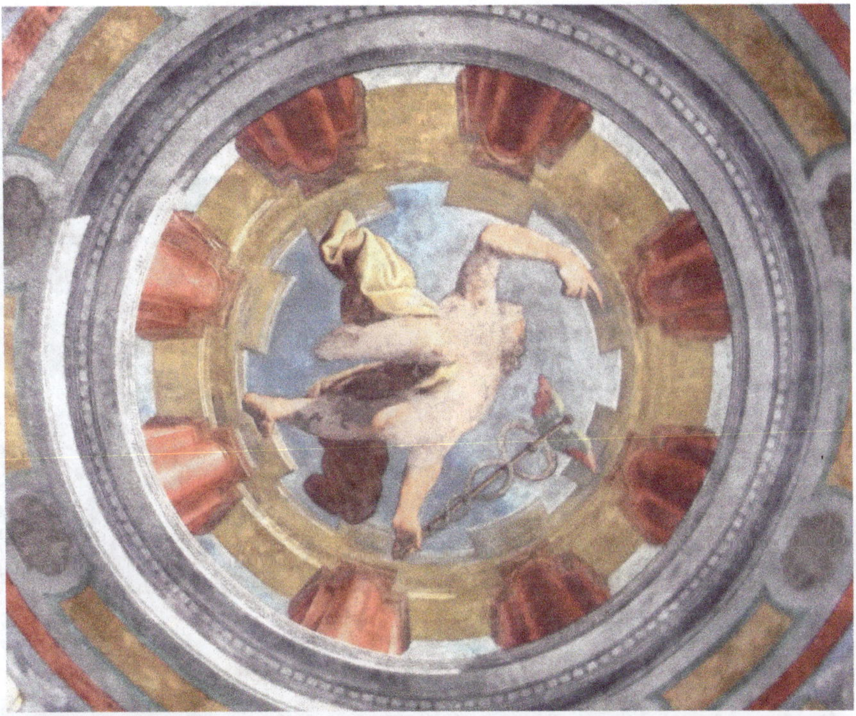

FIGURE 3.1 *Camillo Procaccini and workshop,* Mercury, *1587–1589, Lainate, Villa Borromeo*

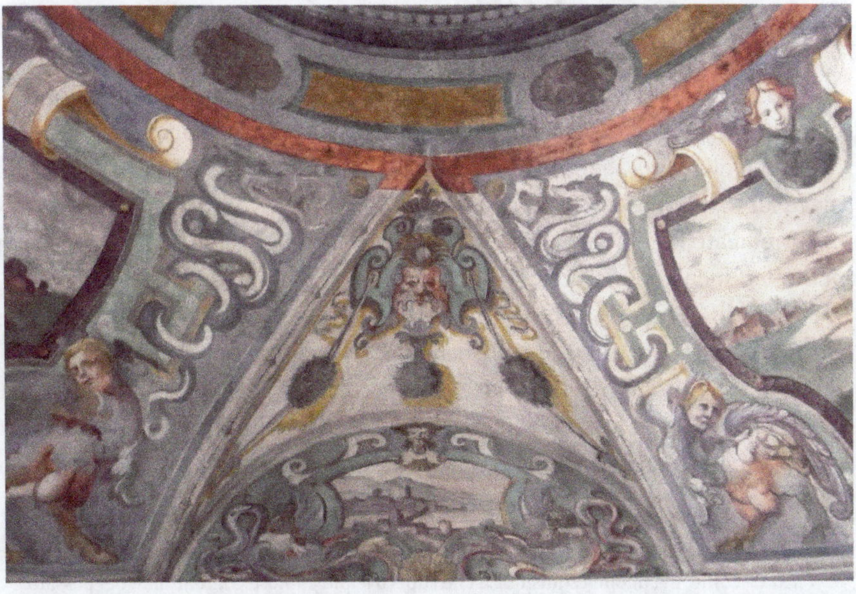

FIGURE 3.2 *Camillo Procaccini and workshop,* Frescoes with animal decorations, *1587–1589, Lainate, Villa Borromeo*

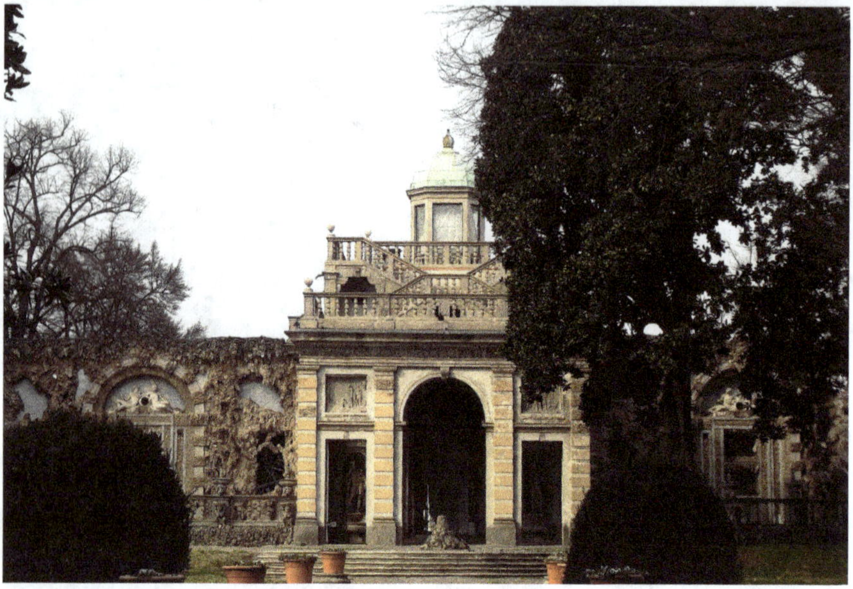

FIGURE 3.3 Nimpheo, *Lainate, Villa Borromeo*

they overcame the most abundant collections of the past and of the present century."[6]

Borromeo's collection was probably displayed in the Nimpheo, completed with water effects, shells, rocks and stones, which were all part of the rich *scenografia*. The architecture itself was also adorned with statues of nymphs and naiads, and it embedded shells, rocks and mosaics (fig. 3.4 and 3.5).

All the elements seem to be part of Borromeo's project to build a specific place for displaying the wonders of nature, or rather, for displaiyng the artificially recreated wonders of nature which the *ingenium* of men could transform and reproduce from nature itself, creating a place of marvel and delight.

Gherardo Borgogni in his book *La Fonte del Diporto* (1598) describes the Nimpheo by stating that "Your Excellency gives a home to the Muses, giving them that Fountain, where in a beautiful contest between Nature and Art everybody enjoys at the same time pleasures and comforts and they draw great things in your name."[7] This passage hints that the place inspired visitors to

6 Borsieri, *Supplimento*, quoted in Morandotti, p. 230.: "Picciola quantità ne raccolse già il Conte Pirro Visconte, ma di tale esquisitezza che superò le più copiose guardarobbe del passato."

7 Gherardo Borgogni, *La Fonte del Diporto* (Bergamo: Comin Ventura e Compagni, 1598), dedication to Pirro Visconti Borromeo.

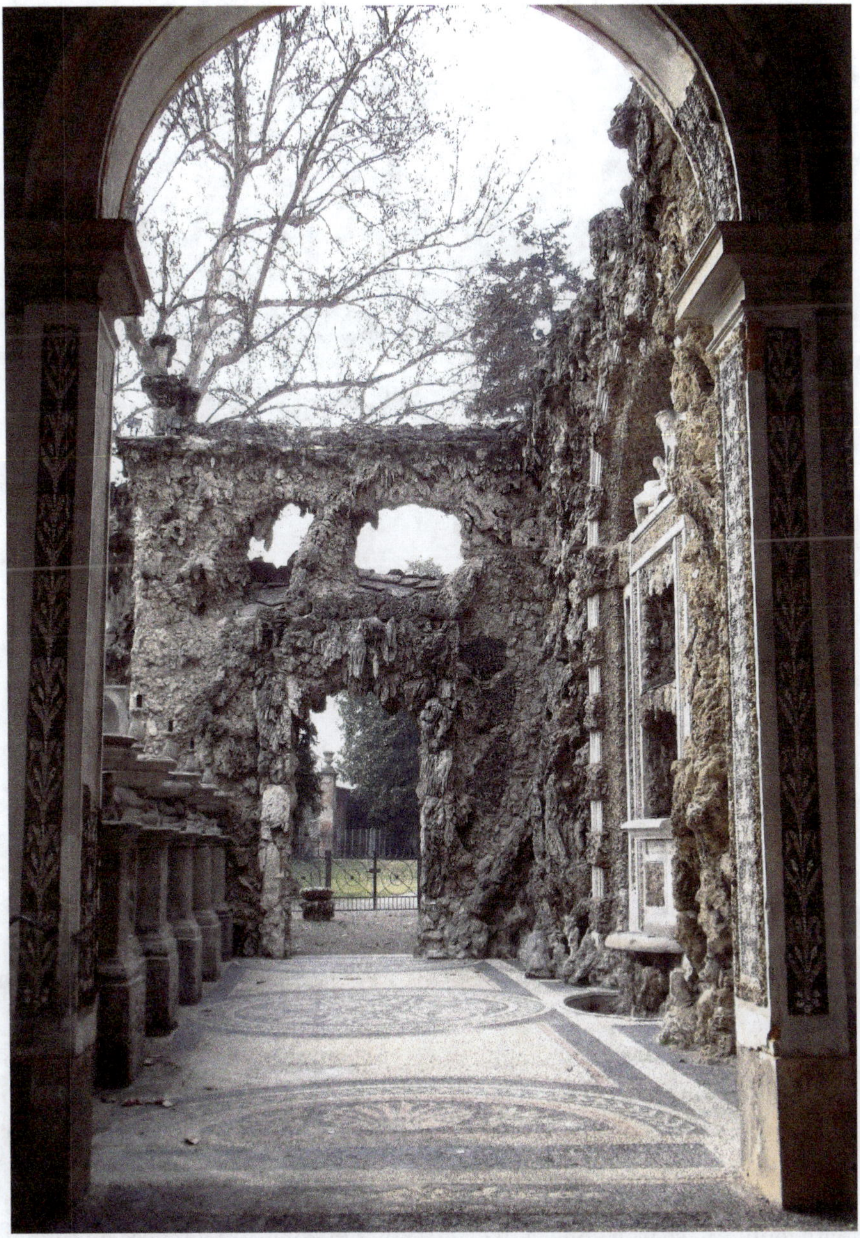

FIGURE 3.4 Nimpheo, *Lainate, Villa Borromeo*

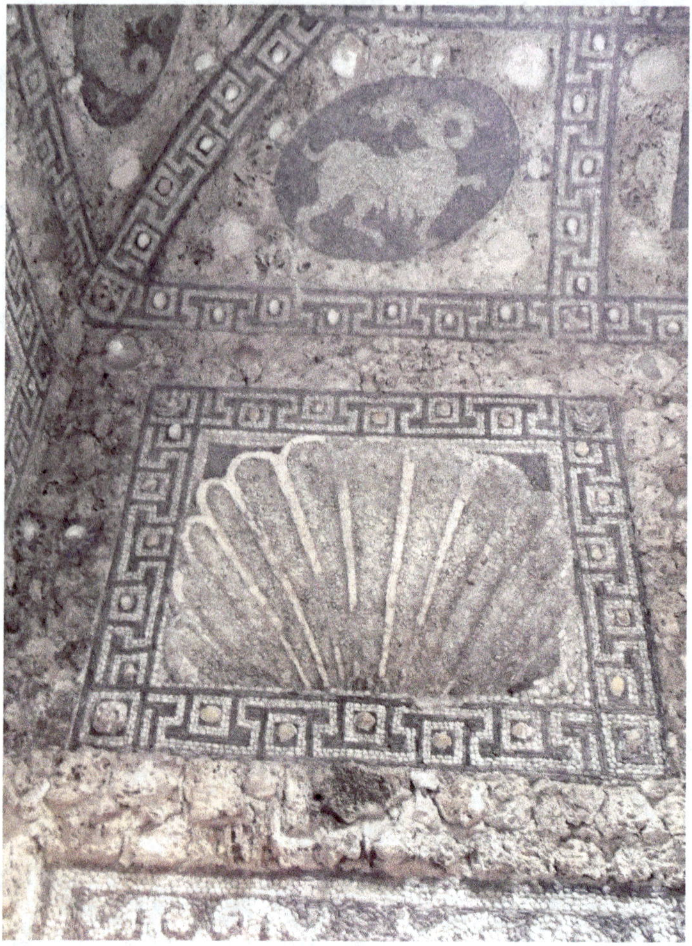

FIGURE 3.5 Nimpheo, mosaic decorations, *c. 1587, Lainate, Villa Borromeo*

"draw things" in Borromeo's name. He is probably referring to the visits Pirro received not only from noblemen and men of letters but also from artists and painters, such as Federico Zuccari or Luigi Scaramuccia, who both recall their stays at the villa of Lainate in their writings. Zuccari states: "We went towards Milan to see the places around, such as the Garden of Count Pirro, adorned with fountains so well done with *artificio* that in Rome or Florence you cannot find more beautiful ones, nor more lavishly adorned with statues of modern sculptors."[8]

8 Federico Zuccari, *Il Passaggio per l'Italia con la Dimora di Parma del Cavalier Federico Zuccaro* (Rome: Tip. Delle Mantellate, 1893), p. 35.

Scaramuccia in his *Finezze dei Pennelli Italiani* (1674) will add:

> They stopped (*Girupeno alias Perugino and Raphael*) to pass the night in those troubled lands where in the morning they could see the beautiful Villa viscontea, which among its pleasures holds beautiful paintings by Camillo Procaccini, placed around the fountain.[9]

This passage hints that not only statues but also paintings were shown around the main fountains of the building, probably only on the occasion of special visits, and almost certainly temporarily.

In the *Idea dell'Architettura Universale* (Venice 1615), Vincenzo Scamozzi records how "honorable and worthy of praise were the fountains of the Villa", distributed throughout the lodges, the arcades and the external rooms as well as the grottos, covered by pebbles or other materials "in different shapes of Nature" (*in varie forme della Natura*), and then adorned with statues made out of stone or metal.[10]

It is quite certain that the villa was overall conceived of as a place to entertain and amaze its visitors, and at the time, it gained a reputation among the noblemen of Milan and beyond. The hospitality and generosity of the count is quoted by different sources, such as the *Relazione della Città di Milano* by Galeazzo Gualdo Priorato, where the villa of Pirro is described as having been visited by the greatest rulers of Italy and other territories:

> No prince or great ruler comes to Milan without going, driven by curiosity, to see that beautiful place, nor any departs from there without remaining satisfied and amazed also by the zeal with which the servants of the count attend to him, having the Villa always been a guest house for Princes and Noblemen. The queen of Spain, the Cardinal as a child, The Dukes of Mantua and other princes all stopped there more than once, and they were all received by the generous Count Pirro.[11]

Guests were obviously not hosted in the Nimpheo but in the rooms of the adjacent Villa, where they could admire the walls and the ceilings painted with

9 Luigi Scaramuccia, *Le Finezze dei Pennelli Italiani* (Pavia: Magni, 1674), p. 150.

10 Vincenzo Scamozzi, *Idea dell'Architettura Universale* (Venice: Girolamo Abrizzi, 1714) part I, book III, p. 343.

11 Galeazzo Gualdo Priorato, *Relazione della Città e dello Stato di Milano* (Milan: Lodovico Monza, 1666), p. 51.

grottesche and decorations, some of which were taken from the natural world while others portrayed fictitious creatures.

Pirro's quest to amaze and entertain his guests is hardly surprising, given his passion for bonfires and fireworks during festivals and special occasions. Morigia testified to this, for instance, in the description of the fireworks that were displayed at the count's Brera palace in December 1590, to celebrate the election of Pope Gregorio XIV—fireworks that were made by the architect Battista Astrologo and that were "amazing and beautiful to see."[12]

Pirro, with the renovations to the Villa, wanted to not only gather a collection, but to create an overall "visitor's experience" for his guests. By both collecting items from the natural world (such as the shells present in his caves and grottos) or recreating them (in the frescoes and many mosaics of natural forms) he used nature for his own ends, playing with it in a way similar to how the members of Blenio were playing with words.

The spaces of the Nimpheo were specifically created and symmetrically arranged in order to host Pirro's collection of *artificialia* and *naturalia*. They showed not only the possibility of recreating the wonders of nature, but also how skillfully the great *ingenium* of artists and artisans could overtake and improve Nature itself. The project of the fountains' system was attributed to the famous idraulic engineer Agostino Ramelli, who already worked in Piacenza. Agostino was the author of the practical book "*Le Diverse et Artificiose Machine*", where he illustrated more than one-hundred machines inspired by Leonardo's studies.[13] The water games and the fountains, as well of the rest of the decorations, were intended as proof of the ability of man to transfer the power of the elements (in this case, of water) into the creation of something beautiful and rare. In Pirro's view, nature was not only collected, but tamed and restrained. In the villa, it was mounted into a rational architectural scheme which staged a fictional wild setting inhabited by Nymphs and Naiads.

A famous poem by Lomazzo in his *Rime*, written in the dialect used by the members of the Academy, summarizes effectively the desire of "taming nature" and reproducing it in artificial ways:

> From the Caves Echo answers politely
> What does Lomazzo do in the Grotto?
> Echo: he is killing serpents, dragons, snakes and sea monsters

12 Morandotti, *Milano Profana*, pp. 26–27.

13 Agostino Ramelli, *Le Diverse et Artificiose Machine* (Paris: in Casa dell'Auttore, 1588).

(…) painted in frescoes, oil, or tempera
and insides he put out of the Grotto Sybils.[14]

As Morandotti noted, it is probably not a coincidence that the book itself was entitled "Rhymes imitating the Grotesques", and that the renovation works at the Villa were taking place during the same years that these poems (1587) and the collection of poems written by the members of the Academy of Blenio entitled the *Rabisch* (1589) were published. Morandotti also states that Lomazzo was probably present on the 3rd of April 1589, when the Villa opened to the public, and that he recited poems praising the count and toasting to Bacchus (patron of the academy and reproduced in more than one statue of the villa).[15] It can hardly be doubted that Lomazzo had a role as counsellor and advisor for the project of the Villa, also given the fact that he personally knew both the patron and the architect designated to renovate it, the Milanese Martino Bassi. The building was finished in 1589 and the collection displayed to a public of men of letters and artists, the heterogeneity of which was in tune with the kinds of members admitted in the academy of Blenio.

The Inventories

There are two inventories of Borromeo's properties in the Villa: one from the 1604 and one dated 1611.[16] Within document A of the 1604 inventory, the list of paintings recorded there have been, to some extent, identified, such as the

14 Lomazzo, *Rime*, p. 500:
"Que fá int' i Grott' ol nost' Omazz? ECHO, Amazza
Bisc, Dragh', Ghizz', Serp', Monster marin, Al rinó
Tanch Asn in roz van al Morin? Al rino-
mèra i gogò che fan i spegaz, e i strazza.
Depenz' a fresch', a oeuiro, a sguazz', e sguazza
Tuch i penij in l'hamorin, corì nò
Ve trighè al son del Tamborin, aruì mò
Sto scatorin, ch'hasto sollaz; sollazza
Drent i Sibill tran feur d'i Grott el Grott'Orb
Luis da Brenta o Basciron, Lyron
De tutt'accord, come on franguel in gabbia.
El bon Zan pavel a gattorb, che nò i Corb
Ma stratonà fa i Cins' int' el Niron,
E i ten per piasè viù in la Vichiabbia."

15 See Morandotti, *Milano Profana*, pp. 15–16.

16 See appendices 5 and 6.

Virgin with Child by Boltraffio or the *Mystical Marriage of Saint Catherine*, now both at the Poldi Pezzoli Museum in Milan. The second part of the inventory is quite intriguing, as it shows different types of objects collected by Borromeo which were destined for the rooms of the Nimpheo –which we know through the pact dated 1611. We have a certain number of medals, a *reliquiario*, clocks, and different cups made by semi-precious stones such as the *lapis armenio*, that is the *azzurrite*. These objects are then listed again in 1611, and some are added, such as a coral fountain, a piece of Indian palm, mushrooms, mother of pearl mounted in silver, a shell mounted in silver, two mother of pearl shells mounted in gold and a so-called oriental jasper mounted in gold. It is said explicitly that all these items were meant to be displayed in the nimpheo, to show the craftsmanship of the artisans who transformed nature in precious objects, creating a prototype of a *wunderkammer* years before the more famous Museo Settala would be established in Milan.[17]

Pirro's interest in not only paintings but also stones and minerals, testified to his knowledge of the artisanal workshops that were transforming these materials, and was probably also shared by Lomazzo and by the members of Blenio (to remain on the examples of the *lapis armenus*, we know that it was also one of the materials for painters to prepare blue, a cheaper alternative to the more expensive lapis lazuli). His expertise regarding precious stones is further indicated by the fact that he was later in life chosen by the Duke of Mantua as a counselor in buying works of art. In fact, we have an interesting exchange of letters between him and Vincenzo I of Gonzaga—one dated 26th of February and the other 16th March 1598—in which we gather that the duke asked Pirro to purchase some rock crystals, and that he chose him as a mediator for the

17 On sixteenth -and seventeenth-centuries collections of *naturalia* and *artificialia* in the so-called "wunderkammern" or cabinets of curiosities see for instance Paula Findlen, *Possessing Nature: Museums, Collecting and Scientific Culture in Early Modern Italy* (Berkeley, Los Angeles and London: University of California Press, 1994); Oliver Impey and Arthur McGregor, *The Origins of Museums: The Cabinet of Curiosities in Sixteenth- and Seventeenth-Century Europe* (Cornwall: House of Stratus, 2001); Marlise Rijks, 'Defenders of the Image. Painted Collectors' Cabinets and the Display of Display in Counter-Reformation Antwerp', in H. Perry Chapman, F. Scholten, and J. Woodall (eds.), *Arts of Display. Netherlands Yearbook for History of Art*, pp. 54–83. See also the on-going researches of scholars such as Claudia Swan (on collections of *Naturalia* and medical collections in Early Modern Holland), Sven Dupré and Christine Göttler (on the collection of Emmanuel Ximenez in Antwerp). More specifically on the cabinet of curiosities of Manfredo Settala, see Marco Navoni, "L'Ambrosiana e il Museo Settala," in *Storia dell'Ambrosiana* (Milan: Cariplo, 2000), pp. 205–255 and Antonio Aimi, Vincenzo De Michele and Alessandro Morandotti (eds.), *Musaeum Septalianum. Una Collezione Scientifica nella Milano del Seicento* (Milan: Museo Civico di Storia Naturale, 1984).

sale of some jewels belonging to the Sala Countess, as another letter dated 3 March 1599 testifies.[18]

The collection in Pirro's Villa of Lainate shows a relatively small but meaningful selection of works of art and decorative arts, along with natural objects and religious items, which testify to the count's particular care not only for the objects themselves but for the context in which they were displayed. The public was supposed to admire the marvels of this world that he created. Lomazzo was most likely involved in the conception of the decorative program, as the formal correspondences between the written arabesques and the grotesque decorations realized in the building are too strong to be ignored. The Lainate Villa combined the grottos with a collection of paintings, among them the divine infant savior by Aurelio Luini, who was, as we have seen, another member of Blenio. By mixing the two art worlds, the noble genre of sacred painting and the grotesque decorations, it became the place to visit and to study both for young artists and men of letters, years before Federico Borromeo's Ambrosiana academy, which was dedicated to the study of painting, sculpture and architecture, was founded in the seventeenth century.[19] This Villa constitutes a visual compedium of the artistic climate which had Lomazzo and his circle of artists as actors and promoters. Along with other places such as the *Scuola di San Luca* or the *Accademia di Blenio*, the Lainate Villa should be counted among the Milanese spaces of artistic and cultural exchanges which we have outlined in this part of the work. Now the second part will focus on how this context and spaces of discussions influenced the making of Lomazzo's *Trattato dell'Arte*.

---

18 For a discussion on these inventories (which concerns especially the paintings and their identification) see Morandotti, *Milano Profana*, pp. 230–243. The inventories are listed in Alessandro Morandotti, "Nuove Tracce per il Tardo Rinascimento Italiano: Il Ninfeo-Museo della Villa Borromeo, Visconti Borromeo, Litta, Toselli di Lainate," *Annali della Scuola Normale Superiore di Pisa*, Classe di Lettere e Filosofia, *15*: 175–185, the *Appendice Documentaria*.

19 Federico, Carlo Borromeo's cousin, became Archbishop of Milan in 1595. He was a fervid collector of art, and he was especially interested in Flemish art (landscape paintings). He was patron of the *Accademia di San Luca* in Rome, which likely inspired him to establish a similar academy for artists in the city of Milan. He was probably in contact with Pirro Visconti, and aware of the project for his villa in Lainate.

PART 2

Color, Perspective and Anatomy

On Movement

More precisely called "On the site, position, decorum, movement, rage and grace of figures", this book consists of thirty-three chapters, in which the concept of movement includes the so-called *moti dell'animo*, that is, the passions of the mind. From chapter I to chapter VI Lomazzo underlines the necessity and the efficacy of movement in painting, arguably taking the concepts on motion from Varchi, Leonardo and Alberti's treatises.[10] In chapter IV ("How the body changes according to the passions of the soul") the concepts are mostly taken from Agrippa's *De Occulta Philosophia*.[11] Then, from chapter XI until the end he demonstrates the movement of different elements and passions (melancholy, shyness, evil, avarice, envy, anxiety, justice, devotion, honor, hatred, magnanimity among others).

On Color

This shorter book consists of nineteen chapters. The term *colore* is used throughout the writing to indicate both natural and artificial pigments, their "friendships" and their "enmities."[12]

On Light

This book consists of twenty-five chapters. His definition of *lume* (ch. III) comprises different (mostly religious) meanings, as he states:

> This word light has different meanings: first and principally it signifies the image of that divine nature which is the Son of God, and the brightness thereof which the Platonics called the image of the divine mind. Secondly, it is the fire of the Holy Ghost. This divine virtue is diffused through all the creatures, in the rational mind becomes the divine grace, and in all living creature is the virtue that preserves and defends, as it is the one, according to Dionisus, of the Seraphins.[13]

10 See Ciardi, *Scritti*, II, pp. 95–103.

11 Ackerman, *The Structure*, p. 74.

12 See following chapter.

13 Lomazzo, *Trattato*, p. 214: "Questa parola di lume si pigli in diversi modi, et significationi. Prima et principalmente significa l'immagine de la divina mente che è il figliuol d'Iddio,

Lomazzo then conveys technical information distinguishing between *lume diretto, riflesso, rifratto,* both natural and artificial, a distinction that could be found in optical treatises of the time such as Witelo's and Alhazen, but that, as we shall see, he could also have taken from more direct conversations with the members of the Milanese academy of the *Val di Blenio.*[14] Subsequently, he explains the effects of lights on various bodies.[15] The last three chapters are dedicated to shadows, according to the three different views (*anottica, ottica, catottica*).[16]

On Perspective

Possibly the most complex book of the Treatise, it consists of twenty-four chapters. After the division and definition of this part of painting, Lomazzo explains the virtues of perspective. He talks about the eye (chapter VII, *Dell'occhio istromento del vedere i raggi*), the distance and the object of the sight. He then lists three major types of perspective (which he defines again *anottica, ottica, catottica*). In the final chapters, he takes Bramantino's perspective as an example worthy of praise (ch. XXI–XXIV *Della prospettiva in generale secondo Bramantino,* and *Primo, secondo e terzo modo della prospettiva di Bramantino*).[17]

According to Ciardi, the value Lomazzo gives to perspective falls beyond the technique: he considers its "moral" value as well, and perspective becomes part of the *decorum* a painting should possess.[18]

---

et unico splendor suo, il quale chiamavano i Platonici, immagine della divina mente. Significa anchor l'ardore dello Spirito Santo. Pigliasi per una virtù divina diffusa nelle creature che nella razionale è la sua divina gratia, et in tutte le creature insieme è la virtù conservatrice, e diffonditrice, come è quello, secondo Dioniso, dei Serafini." Ciardi convincingly argues that this part as well is taken from Agrippa, *De Occulta Phil.*, I, 49.

14 See Lomazzo, *Trattato,* book IV, chapters IV–XI. See the chapter on perspective of this book.

15 Lomazzo, *Trattato,* book IV, chapters XII–XXII.

16 Lomazzo, *Trattato,* book IV, chapters XXIII, XIV, XV.

17 On this issue see Erwin Panofsky, *The Codex Huygens and Leonardo da Vinci's Art Theory* (London: Studies of the Warburg Institute, 1940), pp. 101 f.: "To understand Lomazzo's terms (*in the Treatise*) we must briefly consider the rather complicated structure of his Book on Perspective. (…) He gives the *diffinitione* of the whole subject by dividing it into three parts: 1. *Ottica* (in the wider sense) meaning theory of vision in general; 2. *Sciografica,* meaning theory of light and shade (…). 3. *Specularia* meaning theory of reflection."

18 Ciardi, *Scritti,* I, LVXII.

The first five books of the treatise, characterized by an outstanding amount of direct and indirect sources, and by the author's desire to show his erudition to the reader, were mainly written for one declared purpose: to demonstrate that painting, not only the appreciation of it but the actual practice and exercise of it, had the right to be counted among the "liberal arts." This statement is clearly specified in several parts of the book, first of all in the letter to Don Carlo Emanuele of Savoy, to whom the treatise is dedicated:

> Most noble Sir, these books that are now published on painting (are) under your egregious name, (...) as magnificent protector of the liberal arts. Among which I have no doubt that painting, not much the theory and the contemplation, but the practice and the exercise of it, must have its place.[19]

Notably, finding a place for painting as a "liberal art" was a central issue during the course of the sixteenth and seventeenth centuries in Italy. Although, from Varchi onwards, we have sources on this debate between and among learned artists and men of letters, the real impact on painters' work in studio practice has yet to be defined.[20] Lomazzo then adds a sixth and a seventh book, *Della Prattica della Pittura* and *Della Istoria di Pittura*, which he declares to be explicitly dedicated to painters.

In the book, the artist clearly distinguishes practical knowledge from theoretical knowledge, although this is more of a formal distinction than a substantial one, since in each book of the Treatise most chapters are dedicated to practical examples and explanations of the rules. The author considers the two forms of knowledge complementary and necessary for painters, as he clearly states not only in the Treatise but also, as we have seen in the first part of the work, at the beginning of the eighth chapter of his *Idea* (*Delle scienze necessarie al pittore* - "On the sciences necessary to the painter"). After the distinction between the "practical" and "theoretical" approaches to learn painting, he lists the "sciences" necessary to the painter and the books which the painter must use

---

19 Ciardi, *Scritti*, I, p. 9: "Escono, Serenissimo signore, questi miei libri della pittura sotto il glorioso nome di V.A. come si singolare e fra tutti i prencipi di questa età liberalissimo protettore di tutte l'arti liberali. Fra le quali non ho dubbio, che la pittura, non pur quanto alla teorica et contemplatione, ma anco quanto alla pratica et esercitazione degnissima, non sia da essere annoverata." (Lomazzo, *Trattato*, folio 1).

20 Among the recent studies on this issue, see Francis Ames-Lewis, *The Intellectual Life of the Renaissance Artist*.

to know the *storie*: "The holy scriptures, books on mathematics, hieroglyphics, *storie*, architecture, anatomy, and many other sciences and arts."[21]

In this part my research focuses on three main aspects of the painter's discourse, that is color, perspective and anatomy, with the aims of:

1. Understanding Lomazzo's sources of information (whether direct or indirect);
2. Evaluating the extent to which the artist's knowledge derived from the network of people with whom he was in close contact;
3. Evaluating how the theoretical discourse on color, perspective and anatomy is integrated with more technical information, and viceversa.

21 Lomazzo, *Idea*, p. 81.

CHAPTER 4

Lomazzo's Colors

Lomazzo lists color as one of the principal parts of painting, along with proportion, movement, light and perspective, which he defines as "the differences that make painting a particular art, an art different from all the other arts in the world."[1]

The third book of his *Trattato,* entitled *Del Colore,* is dedicated to the subject and it is a mixture of a theory of colors and a discourse on the practical application of colors in painting.[2] Whilst the first chapters present a theoretical discussion on what color is, where it comes from and what its "friendships and enmities" are, the subsequent chapters display a collection of recipes and a list of the principal colors (which are for him black, white, red, *pavonazzo*, yellow, green, turquoise, ordered according to the hues). Yet, as we shall see also for the part on perspective, the book dedicated to color is not the only place where Lomazzo talks about the subject. Despite the author's strict categorization and division of the different parts of painting at the beginning of the treatise, references and descriptions of the qualities of color, as well as explanations of the techniques of coloring are found in many other parts of his writings. These include the book of light, the book of practice, and the *Idea del Tempio della Pittura*, which was published later but had a close connection with the writing process of the Treatise.[3] Here Lomazzo mentions the subject in three very different chapters, *Delle sette parti del colore* (ch. XIII) – in which he describes the ways of coloring of his *governatori dell'arte, Del modo di colorare i corpi* (ch. XXVIII) and especially in *Della terza parte della pittura et de i suoi generi* (ch. XXI), where he analyzes in more detail the six ways of paintings: "*a oglio,*

1 Ciardi, *Scritti*, II, p. 23: "E perchè le differenze che fanno che la pittura sia arte particolare e sia differente da tutte le altre arti del mondo sono cinque, cioè proporzione, moto, colore, lume, perspettiva, tratterò di ciascuna di queste differenze separatamente in un libro per ordine." (Lomazzo, *Trattato*, p. 15).

2 For a discussion of Lomazzo's theory of colors, see Kemp, *The Science of Art*, pp. 269–274, and more recently Thijs Weststeijn, *The Visible World* (Amsterdam: Amsterdam University Press, 2008), pp. 225–228. On Lomazzo's theory of colors essentials are the studies by Erwin Panofsky, *Idea: ein Beitrag zur Begriffsgeschichte der älteren Kunsttheorie* (Leipzig and Berlin: Studien der Bibliothek Warburg, 1924), Robert Klein, "Giudizio e Gusto dans la Théorie de l'Art au Cinquecento," *Rinascimento*, 1961, *XII*: 105–116 and Moshe Barasch, *Light and Color in the Italian Renaissance Theory of Art* (New York: New York University Press, 1978).

3 Ackerman, *The Structure*, p. 323.

 | DOI 10.1163/9789004330269_007

fresco, tempera, chiaro et scuro, ombrando e lineando solamente", and gives recipes for the mixing and the application of colors. The occurrence of these scattered chapters can be explained by considering the fact (which we mentioned in the introduction) that Lomazzo's actual discussion of the parts of painting does not respect the strict categorization that he makes at the beginning of the whole treatise. Yet despite this fact, book III of the *Trattato* is the part devoted to a systematized discourse on the subject, and it is the one to which we will devote particular attention. The overall organization of the book *Del Colore*, along with other chapters on colors and manners of coloring, gives us a clear indication that by the time the artist published his *Trattato*, this topic had evolved into a literary subject and had become part of a discourse beyond the workshop's walls, in environments such as the *Accademia di Blenio* analyzed in the first part. The structure and arrangement of the discourse on color in the *Trattato* differs completely from the dialogue form adopted by some previous writers on art, such as the above-mentioned Pino in his *Dialogo* or Lodovico Dolce in his *Dialogo dei Colori* (1565).[4] At a first glance, this organization may be consistent with the author's declaration to make a book that would be "useful for painters" in which both the theory and practice of painting obtain the proper place in the book, as the author states in several parts of his writings.[5]

Theoretical notions on colors are at many points appropriated from previous works, and often follow the Aristotelian classification of colors and their properties. These parts are sometimes included with elaboration and mediation by the author, or in some cases without further development. Lomazzo is the first writer on art to build a complex system of colors in connection with planets, elements and emotions.[6] As we shall see, he does not explain these connections in the book *Del Colore*, until, oddly enough, later in the Treatise in the part dedicated to the practice of painting. On the other hand, one can find recipes and suggestions on coloring in book III, most of which are collected and displayed in the few initial chapters. They seem to reflect the author's tendency to include practical notions and painterly materials acquired in the *bottega*, although these recipes are for the most part out of date since they contain a lot of material on how to prepare pigments for illuminations. They are most likely appropriated from a previous source circulating in Northern Italy at the time, and they should be regarded as part of the tradition, developing from Cennini onwards, of collecting and copying recipes for pigments and

4 Pino, *Dialogo* and Lodovico Dolce, *Dialogo dei Colori* (Venice: Giovan Battista Marchio, 1565).

5 See especially the already mentioned letter to Don Carlo Emanuele of Savoy, to whom the book is dedicated.

6 See Klein, *Idea*.

publishing them in writings on art.[7] These recipes are interesting to analyze in detail in order to understand Lomazzo's process of selection and organization of earlier material, his personal additions to these sources, as well as the reasons why he decided to include them in his theoretical discussion on color.

This chapter presents an in-depth analysis of Lomazzo's treatment of *colore* with the aim of understanding both the theoretical discourse that the author presents on this subject throughout the book, and the practical suggestions that he discusses. First, it takes into account Lomazzo's references to color in the Treatise before book III. Secondly, it compares the painter's definition of color with previous definitions in order to understand which one the author actually decides to appropriate. Next, it considers the recipes and the suggestions on techniques of coloring that the author decided to include in the book, giving an overview of other writers on art who decided to include recipes in their theoretical treatises. It then analyzes Lomazzo's terminology regarding coloring and his codification of colors in comparison with previous systems that can be found in some other Italian treatises on art. Finally, it presents the astrological connections that the author makes to build his complex system of colors, showing how some of these connections, particularly between colors and elements, were already made by previous writers on art.

Color in Nature, Color and Nature

Although the definition and the division of color into seven types appear for the first time in book III, a discussion of the functions and usefulness of *colore* is already presented much earlier in the treatise, at the beginning of the first

7 Among various studies on recipes for pigments, see Rutherford J. and George L. Stout Gettens, *Painting Materials: A Short Encyclopaedia* (New York: Dover, 1942); Jonas Gavel, *Color: a Study of its Position in the Art Theory of the Quattro and Cinquecento* (Stocholm: Almqvist & Wiksell International, 1979); Rosamund Harley, *Artists' Pigments: A Study in English Documentary Sources (1600–1835)*, (Washington: Butterworth Scientific, 1997); Jo Kirby, "The Painter's Trade in the Seventeenth Century: Theory and Practice," *National Gallery Technical Bulletin*, 1999, 20:5–49; Arie Wallert (ed. by), *Still Lifes: Techniques and Style: An Examination of paintings from the Rijksmuseum* (Zwolle: Waanders Publishers, 1999); John Gage, *Color and Meaning: Art, Science and Symbolism* (London: Thames & Hudson, 1999). Amongst the most recent publications, see Spike Bucklow, *The Alchemy of Paint: Art, Science and Secrets from the Middle Ages* (London: Marion Buyars, 2009); C.P. Biggam, C.A. Hough, C.J. Kay and D.R. Simmons (eds.), *New Directions in Color Studies* (Amsterdam and Philadelphia: John Benjamins Publishing Company, 2011); S. Panayotova (ed. by), *COLOUR: The Art and Science of Illuminated Manuscripts* (Turnhout: Brepols, 2016); see also the ongoing research project 'Making and Knowing' directed by Professor Pamela Smith at the Columbia University in New York.

book dedicated to the natural and artificial proportion of things. It is embedded in a philosophical discussion on the imitation of nature, quite clearly of Aristotelian derivation. Here, Lomazzo starts a discourse on the "property of painting", which is: "To imitate nature's colors (...) with the help of geometry, arithmetic, perspective and natural philosophy."[8]

As Lomazzo states, "Painting represents things with color similar to that of natural things, and we must consider that being the painter *artifex* he needs to proceed according to Nature, which (as all natural philosophers state), first of all creates the matter and then the form."[9] While he does not seem to question the notion of the superiority of drawing over color, as derived from Vasari, he certainly grants color an important role: "And they (painters) should note that, however skilled in coloring they may be, if they do not possess *disegno* they do not have the matter of painting and consequently they lack the substantial part of it. In any case, one cannot deny that the force of coloring is great indeed."[10]

The analogy between color as form and drawing as matter is further developed in the following paragraphs, when the author infers that color gives images their distinctive traits, the "form", so that one can discern a particular man:

> But when in addition to drawing and the right and equal proportion one adds the proper color, then one gives the ultimate perfection to the figure and makes everybody able to see which man is it, and one can say that it is for instance the emperor Charles V, or his son Philip, or that it is (the color of) the melancholic man, the phlegmatic, the sanguine or the choleric (...) And it is for this distinctive quality, that is to demonstrate to the eye things with similar colors, that painting is an art that differs from all

---

8 Ciardi, *Scritti*, II, p. 26: "Ed è imitatrice e come a dire simia de l'istessa natura, la cui quantità, rilievo e colore sempre cerca di imitare. Il che fa con l'aiuto de la geometria, aritmetica, perspettiva e filosofia naturale, con tanta e così retta ragione che non può essere più." (Lomazzo, *Trattato*, p. 18).

9 Ciardi, *Scritti*, II, p. 30: "La pittura rappresenta le cose con colore simile a le cose naturali. Nel che si ha da considerare che essendo il pittore artefice, ha da procedere secondo il modo de la Natura, la quale prima presuppone (come dicono tutti i filosofi naturali) la materia de le cose e poi gli dà la forma." (Lomazzo, *Trattato*, p. 24).

10 Ibidem: "E perciò avvertiscano che quantunque siano eccellenti e miracolosi in colorire, se non hanno disegno non hanno la materia de la pittura e conseguentemente sono privi de la parte sostanziale di lei. Non si niega però che non sia grandissima la forza del colorire."

> the others, and especially from sculpture, in which it is clear that one does not use colors; from this you deduce the greatness and superiority of painting toward sculpture, since a painter does what a sculptor cannot: he imitates Nature perfectly.[11]

In these dense passages, Lomazzo takes the well-known argument of the *paragone* between sculpture and painting, already developed almost forty years earlier by Benedetto Varchi, and embeds it into a philosophical definition of color as the form, the particularity and the quality of painting. Moreover, he underlines the importance of light for color, a concept which he will further develop in the Book of Light, since "color cannot be seen without light, being the latter the last surface of the opaque and thick body (according to philosophers) so that it is needed for a painter to be excellent in coloring, and he must enquire deeply and acutely on the effects of light when it enlightens color, so that in observing with diligence these effects, he will become great in the art of painting."[12]

Other links to the discourse on light and also issues of *decorum*, that is the proper way of depicting images, were discussed earlier in chapter II, "On the division of painting", where the author exemplifies how to depict the part of the body that is opposite of the sun, while using the right colors:

> Color, together with light, must be considered in two ways, in the natural way and in perspective, such as we have already said for proportion. We call enlightened natural color the one that man has naturally or the thing

11 Ciardi, *Scritti*, II, p. 31: "Ma quando oltre al disegno e quantità proporzionata giusta et uguale aggiunge il color simile, allora dà l'ultima forma e perfezione a la figura e fa sì che ognuno che la vede discerne di qual uomo è e sa dire, per esempio che è de l'imperatore Carlo V, o di Filippo suo figliolo, che è d' uomo melanconico, o flemmatico, di sanguigno o di collerico. (...) E per questa particolarità c'ha in sé la pittura, cioè di dimostrate a l'occhio le cose con colore simile, ella si fa differente da tutte le altre arti, e massime da la scoltura, ne la quale è chiaro che non si adopera colore, d'onde si cava ancora l'eminenza di essa pittura et eccellenza sopra la scoltura; poichè il pittore fa quello che lo scoltore non può perfettamente fare in imitare con l'arte sua la Natura, così come perfettamente l'imita il pittore." (Lomazzo, *Trattato*, p. 25).

12 Ciardi, *Scritti*, II, p. 32: "E perchè il colore non si può vedere senza il lume, non essendo egli altro, secondo i filosofi, che l'ultima superficie del corpo terminato e spesso, allumata, è bisogno che il pittore che vuole essere eccellente coloritore, sia peritissimo e sagacissimo investigatore de gli effetti che fa il lume quando alluma il colore, che così osservando con alta e profonda considerazione questi effetti, diventerà unico ne l'arte della pittura." (Lomazzo, *Trattato*, p. 26).

> that one wants to represent, and we also call this natural, not in the way philosophers mean it, but in the painters' way. So that part of the natural body that is opposite to the sun has three degrees of red, and it receives other three degrees of light from the sun. If a painter wants to represent this part in a natural way, he will put three nuances of red and other three of bright color with which he will depict the light. In so doing he will represent both color and natural light.[13]

This passage is crucial in more than one respect: aside from linking once again color to light, Lomazzo draws for the first time a division between natural and artificial colors, stressing the importance of "natural color" as the one a painter must use in order to imitate nature. Moreover, he points out that his aim in the book is to discuss these matters not as a philosopher, but as a painter. This is a recurrent statement throughout the author's literary achievements, and it should be regarded not so much as an admission of his amateurish literary education (as the *homo senza littere* Leonardo presented himself), but as a specific declaration of intentions in which the author codifies and categorizes painterly knowledge as a set of practical notions and expertise, of which he has first-hand experience, in a manner that is understandable for fellow painters. In the later *Idea*, Lomazzo would go further in expressing the importance of "natural colors", linking the necessity of learning how to use colors following the example of great masters, and appropriating the right manner of coloring in order to represent ideas. He also criticized the manner of coloring of certain Venetian artists, while praising others such as Titian and Giorgione. In chapter XXVIII "On the manner of coloring bodies" he states that:

> Having sufficiently discussed all the methods of coloring in their place, I will not elaborate further upon them here, but I will only remind the painter of one sure piece of advice: he must strive with all his forces to imitate the natural color in everything, whatever the attitude or movement he wants to represent, conforming with what he had conceived in

13 Ciardi, *Scritti*, II, p. 37: "Il colore insieme con la luce si considera parimenti in due modi, naturalmente et in perspettiva, come abbiamo detto de la proporzione. Colore illuminato naturale è quello che ha naturalmente l'uomo o la cosa che si vuole rappresentare, e naturale chiamiamo in questo loco, non secondo lo stretto significato de filosofi, ma al modo de' pittori. Per essempio quella parte del corpo naturale che mira rettamente e sta opposta al sole ha tre gradi di color rosso e riceve altri tre gradi di luce dal sole. Ora, se il pittore vorrà rappresentare questa parte, appunto come ella si vede nel naturale, questo farà ponendo tre gradi di color rosso et altri tre di color chiaro co'l quale esprimerà la luce; e così rappresenterà naturalmente il colore e la luce naturale." (Lomazzo, *Trattato*, p. 31).

> his mind, as clever Titian, Giorgione and other great painters always did. For this reason, their works appear really colored by nature, as if each thing truly represents reality (...) I urge those seeking honor never to make the color chosen seem artificial itself, because that actually weakens the power of *disegno*. This defective manner of coloring is commonly employed by certain Venetians, and although widely appreciated by fools and even masters of this art, it corrupts the practice of their fellow compatriots cited above.[14]

Apart from linking the representations of colors to the images of the mind (which scholars such as Robert Klein discussed thoroughly), this passage, as Chai recently noted, mainly criticizes the application of pure saturated colors without shades or undertones, the "coloring for the sake of coloring" without any reason related to the one aim of imitating nature.[15] Lomazzo then declared that he was "forced to condemn that detestable practice of coloring following colors, which has now so taken hold that Italy and the Germanies have become totally infested with it."[16] This statement testifies to Lomazzo's knowledge of Northern art outside of Italy (which will be further discussed in the following chapter), and it clarifies his position on that manner of coloring he does not consider worth following.[17] However, six years before publishing these statements on how color should and should not be used by painters in the *Idea*, the author already developed a pregnant discourse on the intrinsic properties of color in his book *Del Colore*.

La Estremità Della Cosa Giudicata: Lomazzo's Definition of Color

Del Colore includes nineteen chapters and, second only to the book on movement, which consists of seventeen chapters, is the shortest book of the *Trattato*.[18] It starts with a rhetorical praise concerning the virtues of the subject treated, and is followed by an explanation of the necessity of coloring within the art of painting and a definition of "what color is, how many kinds

14 Lomazzo, *Idea*, ch. XXVIII. Ciardi, *Scritti*, I, p. 323. Translated in Chai, *Idea*, pp. 124–125.

15 See Chai, *Idea*, p. 220 and section VI of the introduction.

16 Lomazzo, *Idea*, chapter XVIII, translated in Chai, *Idea*, p. 125, and Klein, *Idea*, II, p. 605.

17 See chapter 1 in part I, "The author's travels".

18 See appendix 7.

there are and whence colors are caused."[19] For this part of the book the painter predictably refers to the authority of Aristotle, stating that "Aristotle defines Color to be a visible quality which is limited and bounded in the surface or extremity of a dark body."[20] As Ciardi notes, this definition is most likely taken from Lodovico Dolce's *Dialogo dei Colori* (1565), although I believe that Lomazzo mixes the quotations that Dolce took from Aristotle when he describes both *lume* and *colore*.[21] While quoting Aristotle's *De Sensu et Sensibilis,* Dolce states that:

> Pythagoreans believed that color was nothing but surface. Plato in his Timaeus said that it was light. Aristotle is somewhere in between: color was the boundary of the body, not of that part in which the body is contained, which would be the surface, as Pythagoreans say, but of the *lucidezza*, however not at the boundary, which would then be light, as Plato said. Color therefore is the extremity and the boundary of the lucid body.[22]

And, few lines below:

> Light, as Aristotle says, is a visible quality, which the opaque body, that is the shaded part which is illuminated by the enlightened body (…) Therefore, to see colors one must seek the medium and the light. One must look for it through the medium rather than the colors. Therefore,

19 Lomazzo, *Trattato*, book III, chapter III. On his praise of colors see Lomazzo, *Trattato*, p. 190: "E questo vanto che si può dare in questa parte alla pittura, io giudico che sia uno dei maggiori e più illustri che si possa dare ad arte alcuna. Oltre che tanto più questa s'inalza sopra le altre e risplende, quanto che per gl'occhi, principal senso, opera e rappresenta la bellezza e tutte le cose conforme a quanto creò giamai Dio."

20 Ciardi, *Scritti*, p. 167: "Che cosa sia il colore e le sue spezie e d'onde si cagionino i colori" (Lomazzo, *Trattato*, p. 190).

21 Dolce, *Dialogo dei Colori*. On Dolce sees among others R.H. Terpening, *Lodovico Dolce: Renaissance Man of Letters* (Toronto: University of Toronto Press, 1997).

22 Dolce, *Dialogo dei Colori*, p. 7: "I Pitagorici credettero il colore altro non esser che superficie. Ma Platone nel suo Timeo disse, lui esser lume. Egli è vero che Aristotele, tenendo una strada di mezzo, stimò che'l colore fosse termine di corpo, non di quella parte, da cui è contenuto esso corpo, che questo sarebbe superficie, come vogliono i Pitagorici: ma della lucidezza, né però non terminata, che ciò sarebbe lume, come piacque a Platone. Colore adunque è termine e estremità di lucido e terminato corpo."

> the thing (that is, color) that cannot be found through the medium, consequently searches the medium and the light, and this is color.[23]

Those last words seem to be the ones Lomazzo uses to make his own definition of color:

> Color, as Aristotle says, is the extremity of the thing we judge or of the thing that is visible in the extremity of a body, that is color is the visible quality limited and bound in the surface or extremity of an opaque body, which before it be lightened, is visible only in potentiality, and by the benefit of the light may be actually seen. So the light in an opaque and dense body causes that color, by the mutual work of the first qualities.[24]

Dolce had already oversimplified (and to some extent misinterpreted) Aristotle's words, and Lomazzo followed his simplification:

> We have already in On the Soul stated of Light that it is the color of the Translucent, (*being so related to it*) incidentally; for whenever a fiery element is in a translucent medium presence there is Light; while the privation of it is Darkness. (...) When the Translucent is in determinate bodies, its bounding extreme must be something real; and that color is just this "something" we are plainly taught by facts -color being actually either at the external limit, or being itself that limit, in bodies. (...) We may define color as the limit of the Translucent in determinately bounded body. For whether we consider the special class of bodies called translucent, as water and such others, or determinate bodies, which appear to

23 Dolce, *Dialogo dei Colori*, p. 8: "Lume, come dice Aristotele, è visibile qualità, la quale riceve il corpo opaco, cioè ombroso, illuminato da corpo lucido (...) Onde ne segue, che per vedere i colori si ricerchi il mezzo e il lume. Il che si ricerca per il mezzo e non per essi colori. Percioché la cosa che non si può vedere, se non per via di mezzo, ricerca esso mezzo e il lume. E tale cosa è il colore." (Ciardi, *Scritti*, II, pp.167–168, n.1).

24 Ciardi, *Scritti*, II, p. 167: "Colore, come dice Aristotile, è là estremità della cosa giudicata o visibile in corpo terminato, overo è la qualità visibile terminata nella estremità del corpo opaco, la quale innanzi che sia allumata, è visibile in Potenza, e per beneficio del lume si vede in atto, percioché il colore è cagionato dalla luce nel corpo opaco e spesso, operando insieme le prime qualità." (Lomazzo, *Trattato*, p. 190).

> possess a fixed color of their own, it is at the exterior bounding surface that all alike exhibit their color.[25]

The definition of color in Lomazzo follows Dolce's Dialogue, and, in fact, as Moshe Barasch already noted, it implies that color belongs to the object and that, at least namely, color is separated from light.[26] On the other hand Lomazzo's division of color into "seven kinds, that is seven manners of colors" could be taken from ch. XXXVI of Cennino Cennini's *Libro dell'Arte*, entitled "*Come ti dimostra i colori naturali; e come dèi macinare il negro*", in which it is stated that:

> You should note that the natural colors are seven: four have an earthly nature, which are black, red, yellow and green. Three are the natural colors that are helped artificially: white, ultramarine blue and light yellow. But let's say no more about this and let's come back to the color black.[27]

In this case, Lomazzo differs in his choice of colors, as he lists: "Seven are the kinds or the way of colors: two are at the extremities and fathers of all the others, these are black and white, and five are intermediate, these are pale, red, purplish and green."[28] In this passage he includes black and white as colors along with the *pallido*, the red, the purplish and the green, forgetting to list the

25 Aristotle, *De Sensu et Sensibilibus*, translated by J.I. Beare, *On Sense and the Sensible* (Adelaide: The University of Adelaide Library, 2014), p. 3.

26 Moshe Barasch, *Light and Color*, pp. 160–161.

27 Cennino Cennini, *Il Libro dell'Arte,* edited by Carlo and Gaetano Milanesi (Florence: Felice Le Monnier, 1859), ch. XXXVI, p. 23: "Sappi che sono sette colori naturali; cioè quattro propri di lor natura terrigna, siccome negro, rosso, giallo e verde: tre sono i colori naturali, ma voglionsi aiutare artifizialmente, come bianco, azzurro oltremarino, o della Magna, e giallorino. Non andiamo più innanzi, e torniamo al nero colore."

On Cennino Cennini see Filippo Baldinucci, *Notizie de' Professori del Disegno* (Florence: D.M. Manni, 1769), pp. 478–487; Among the vast literature on his *Libro dell'Arte*, see Charles Eastlake, *Materials for a History of Oil Painting* (London: Longman, 1847); Roberto Longhi, "Letteratura Artistica e Letteratura Nazionale," in *Paragone*, 1952, 33: 7–14, especially p. 10; Frederick Antal, *La Pittura Fiorentina e il suo Ambiente Sociale nel Trecento e nel Primo Quattrocento* (Turin: Einaudi, 1960); Miklos Boskovits, "Cennino Cennini Pittore non Conformista," *Mitteilungen des Kunsthistorisches Instituts in Florenz*," 1971, 17:201–222; Sandro Baroni and Silvia Bianca Tosatti, "Sul Libro dell'Arte di Cennino Cennini," *Acme: Annali della Facoltà di Lettere e Filosofia dell'Università degli Studi di Milano*, 1998, 51: 52–72.

28 Ciardi, *Scritti*, II, p. 168 "Sette sono le spezie, overo maniere dei colori. Due sono estremi e come padri di tutti gl'altri, e cinque mezzani. Gl'estremi sono il nero et il bianco, e i cinque mezzani sono il pallido, il rosso, il purpureo et il verde." (Lomazzo, *Trattato*, p. 191).

seventh color he considers intermediate. One can safely suppose that he is talking about turquoise, since the author dedicates to this color, as to the other six colors, a chapter of its own. Another inconsistency is that the painter lists the so-called *pallido*, while he dedicates then a chapter to the yellow color and not to "pale." Moreover, the purple or purplish listed in this passage further down in the chapter is mentioned as "*pavonazzo* or *morello*", a color already quoted by Vasari and mentioned by other authors such as Giovan Battista Armenini.[29] Whatever the reason for this change, it certainly speaks to what can be called the editorial difficulties Lomazzo encountered in making the book. Moreover, the author's choice of the principal colors (especially *pavonazzo* and *pallido*) reveals the integration of painters' practical knowledge into theoretical discourse, in a manner similar, as we shall see, to the choice of the terminology for the different parts of the body in discussing human anatomy.[30]

A comparable division of colors (although the number seven is not specified) can be found almost forty years before in Michelangelo Biondo's book *Della Nobilissima Pittura* (1549). Biondo dedicates a chapter of his book, entitled *Di Vari Colori,* to the subject. He defines as "natural colors" white, black, light blue, red, yellow and green, which are "the most necessary colors for painters."[31]

The division into seven colors is also mentioned later in the Treatise, in book VI dedicated to the practice of painting, where Lomazzo quotes again Aristotle, despite the fact that the latter does not divide colors into seven.[32] As we shall see when discussing perspective, the necessity of quoting names of ancient authorities is particularly prominent in the theoretical parts of Lomazzo's definitions of color and its meaning, whereas when he recalls anec-

29 Giorgio Vasari, *Vite* (1568), edited by Rosanna Bettarini and Paola Barocchi (Florence: Sansoni, 1967), p. 167: "E perchè usava Buonamico, per fare l'incarnato più facile, di campeggiare, come si vede in questa opera, per tutto di pavonazzo [e] di sale -il quale fa col tempo una salsedine che si mangia e consuma il bianco e gl'altri colori-, non è maraviglia se quest'opera è guasta e consumata, là dove molte altre che furono fatte molto prima si sono benissimo conservate." Armenini, *De' Veri Precetti*, p. 125: "Ma si vuole certe avvertenze ancora quando si lavora nel porvi alcuni colori, perciocche è come lo smalto ed il pavonazzo."

30 See ch. VI.

31 Michelangelo Biondo, *Della Nobilissima Pittura* (Venice: Bartolomeo l'Imperadore, 1549), chapter XXIV, *Di varii colori*, pp. 20r-21r: "Finalmente osservando io la pittura, per cagion di pittor novelli qui ho terminato di giungere, li più necessari colori di quali consta la pittura, el primo colore gli è bianco, poi negro, azuro, rosso, zalo, et verde, et cotesti colori sono colori naturali, et oltra vi è laca, carmesino, et gli altri colori di sua spetie."

32 See Ciardi, *Scritti*, II, p. 266, n2.

dotes or ancient stories or when he explains how the suitable colors for each figure should be applied in accordance with the principle of *decorum*, he seldom quotes the authors or the books from which he takes information.[33]

Le Materie dei Colori: Lomazzo's Recipes and the Paduan Manuscript

After these opening theoretical chapters of *Del Colore*, Lomazzo continues the discussion on color with more "practical" chapters dedicated to the "matters in which you can find colors."[34] Four years prior, we find a similar discourse on "the Matter of Color" in the *Osservazioni Sopra la Pittura* by Cristoforo Sorte (1580). The beginnings of chapter IV and chapter V of Lomazzo's *Del Colore* bear similarities with Sorte's passage:

> But let us now turn to the materials of colors, and we will start from the general ways of painting, in order to make these things more understandable and so that they can be useful to learn. The ways of painting are four: *acquarello*; that is when colors are used on paper; *guazzo,* that is when these colors are used on canvas; a *fresco* and *a secco*, on wall; and *a oglio*, that is when one wants to paint on board, though sometimes it is used also on wall.[35]

Lomazzo refers to four rather than three ways of painting, omitting watercolor:

33 For instance, at the very beginning of *Del Colore*, there are different mythological episodes taken from Pliny's *Naturalis Historia*, which were already mentioned in other writers on art of the time such as Benedetto Varchi or Paolo Pino, which Lomazzo is more likely to have read.

34 Haydocke translated the title of this chapter simply as "on the matters of colors," but this translation does not convey the concept of color being a visible presence that can be found in various materials.

35 Cristoforo Sorte, *Osservazioni Sopra la Pittura* (Venice: Girolamo Zenaro, 1580), p. 281: "Ma passiamo alla materia dei colori, intorno alla quale, accioché i nostri ragionamenti siano facili et ordinati, e le cose, che ci occorreranno a trattare, meglio siano intese et alcuna utilità se ne possa indi cavare, fa di mestieri che prima sappiamo i modi generalmente di adoperar essi colori li quali modi sono quattro, cioè: ad acquerello, il che aviene quando s'adoperano essi colori su la carta; a guazzo, quando si dipinge in tela; a fresco et a secco, nel muro; et a oglio, il che si fa ordinariamente quando si vuole operar su le tavole, benché anco si faccia a secco nel muro alcuna volta."

> Because some colors in all three ways of painting cannot be used without being destroyed, such as frescoes over quicklime, oil painting and working on tempera, I will distinguish them according to the peculiarities which each of them entails.[36]

A similar distinction was to be made by Armenini in his *De' Veri Precetti della Pittura* two years later, in which he also distinguished between natural and artificial colors:

> I believe every average painter should know that all colors in painting must be of two kinds, either natural (also called *di miniera*) or artificial, and you must dilute them with three liquids: water, glue and oil. The first way is called frescoes, the second *a secco* and the third *a oglio*.[37]

The recipes presented in Lomazzo's chapters on colors, that is chapter IV, VII and VIII, are recipes for pigments that are strikingly similar to those presented in a Paduan manuscript published by Mary Merrifield.[38] Previously, they have been compared to Cennini's recipes in his *Libro dell'Arte*, a comparison, as Ciardi admits, that was not very fruitful, since Lomazzo did not follow Cennini's recipes for the most part.[39] In her commentary, Merrifield states that the author of the manuscript copied from Lomazzo's work, mainly because "The work bears intrinsic evidence of having been composed at a later period."[40] Despite this statement, there is an issue that cannot be ignored: the fly-leaf preceding the main text of the manuscript, written in the same hand-writing and in similarly colored ink, is a sonnet dedicated to the Prince Emanuel

36 Ciardi, *Scritti*, II, p. 169: "Perchè alcuni colori non si possono adoperare senza la morte loro in tutte tre le spezie di dipingere, che sono il fresco sopra la calce fresca, a il lavorar a oglio et il lavorare a tempera, gl'anderò distinguendo secondo che a ciascuna di queste tre maniere di dipingere si convengono e si comportano." (Lomazzo, *Trattato*, p. 192).

37 Armenini, *De' Veri Precetti*, ch. VII–IX, p. 105: "Stimo che gli sia noto ad ogni pittor mediocre che tutti i colori, i quali si adoperano per dipingere, debbano essere di due specie, cioè naturali, che si dicono ancora di miniera, et artificiali, i quali si distemprano a lavorarli communemente con tre liquori, i quali sono acqua, colla et oglio: il primo si chiama lavoro a fresco, l'altro a secco, et il terzo ad oglio."

38 Mary P. Merrifield, *Original Treatises on the Art of Painting* (New York: Dover Publications, 1967), pp. 648–717: "Paduan Manuscript, MS 992: *Ricette per fare ogni sorte di colori*." For a direct comparison of (part of) the two texts see appendix 8.

39 Ciardi, *Scritti*, II, p. 168, n1.

40 Merrifield, *Original Treatises*, II, p. 644.

Philibert of Savoy, who died in 1580. The sonnet, apparently composed by the *canonico* Michelangelo Blanchiardi, would indicate that the manuscript could not have been composed later than Lomazzo's Treatise. On the other hand, the manuscript presents some characteristics that could indicate a later dating. These include the typical seventeenth-century handwriting (see fig. 4.1), and the fact that some ingredients and techniques mentioned in the later part of the text, above all the use of cochineal for led lake, saw a wide diffusion in Italy only in the seventeenth century (although it was introduced in the last part of the sixteenth century). As the manuscript states: "Another sort of fine lake: Take 12 grains of powdered cochineal or *finegrana*, add it to it 2 ounces of lye; leave the infusion about 2 hours; strain it through a linen cloth and put it over hot cinders." (See fig. 4.2).[41]

Also, as Merrifield mentions in her commentary, in this manuscript Gamboge is mentioned, which was introduced in medicine around 1600, and was also used in painting in combination with lemon juice and rock alum.[42]

Despite this evidence, Lomazzo's writing appears more elaborate and extended (see appendix 8), which may indicate that the copying process was done by him, rather than by the manuscript's author. Moreover, Lomazzo's text occasionally presents differences in describing the ingredients, for instance to obtain the color black: *"Negro si fa col 'fumo d'oglio di noce arso"* is copied *"quelli che fanno il nero sono l'avorio arso"* - to burn ivory (*avorio*) instead of nut oil (*oglio di noce*). This could be seen as reinterpretation on behalf of the painter, since to burn bones –not necessarily ivory, which was considered as too precious- was another way of obtaining black.[43]

For these inconsistencies, I tend to believe that the manuscript's author did not copy directly from Lomazzo, or vice versa, but that the two writers must have had a common earlier source, which they both copied, a source that dates from before 1580 (the date of the death of Emanuele Philibert). Given Lomazzo's connection to the court of Turin (as we have seen, the whole treatise is dedicated to the ruler of the city), he might have come across this collection of recipes that he then decided to include in the book.[44] The Paduan manuscript would then be a seventeenth-century compilation of recipes, the first part of

41 Merrifield, *Original Treatises*, II, p. 702.

42 Merrifield, *Original Treatises*, pp. 644–645. For a recent discussion on medical recipes, see Elaine Leong, "Collecting Knowledge for the Family: Recipes, Gender and Practical Knowledge in the Early Modern English Household," *Centaurus*, 2013, *55*: 81–103.

43 For some examples on the vast literature on pigments and color-making see note 195.

44 The connections between Lomazzo and Turin have already been discussed in the first part of the book (see "The Author's Travels").

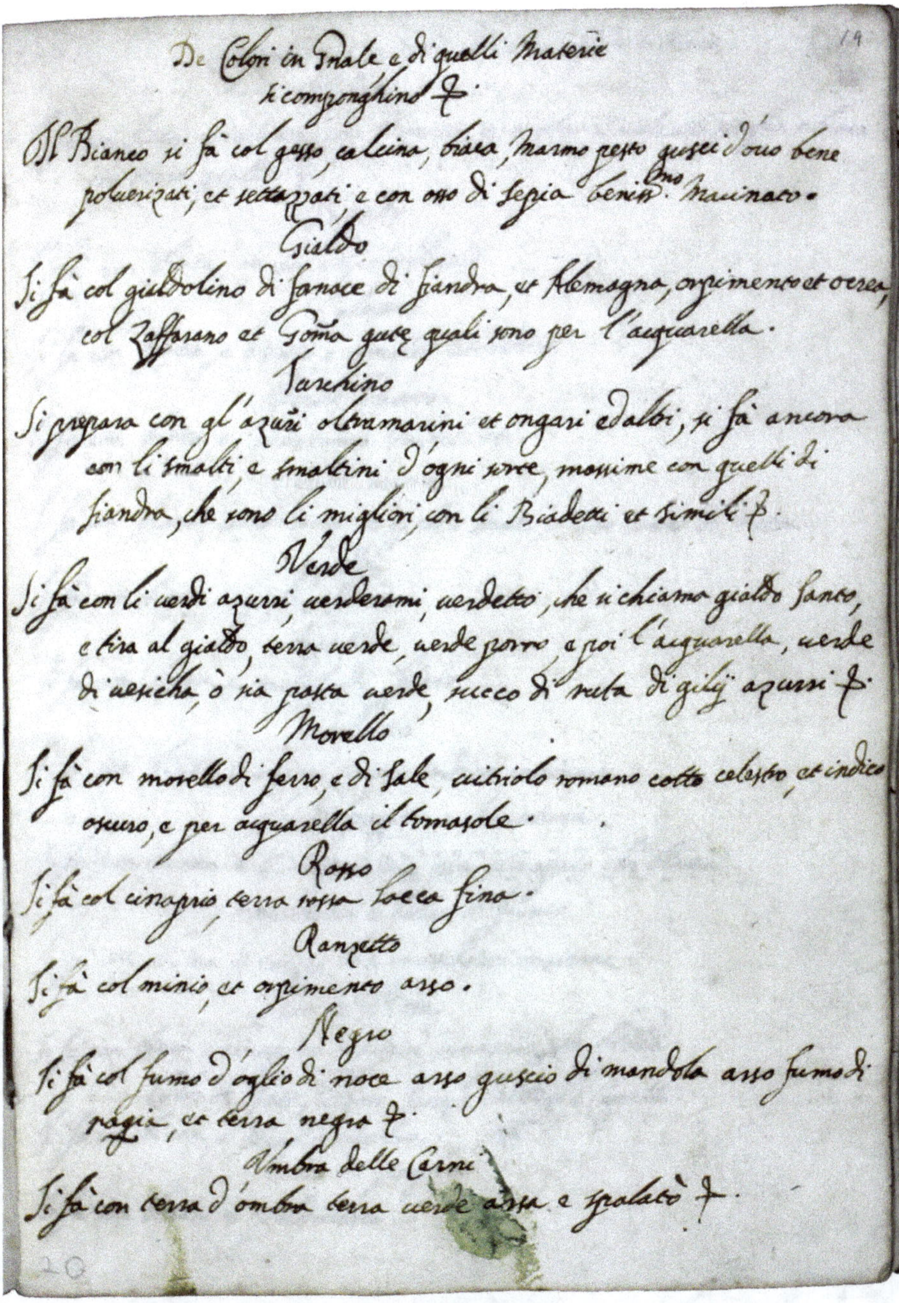

FIGURE 4.1 *Ms 992,* Recipes for all sorts of colors, *fol. 19, c. 1600–1650, Padua, Biblioteca Universitaria*

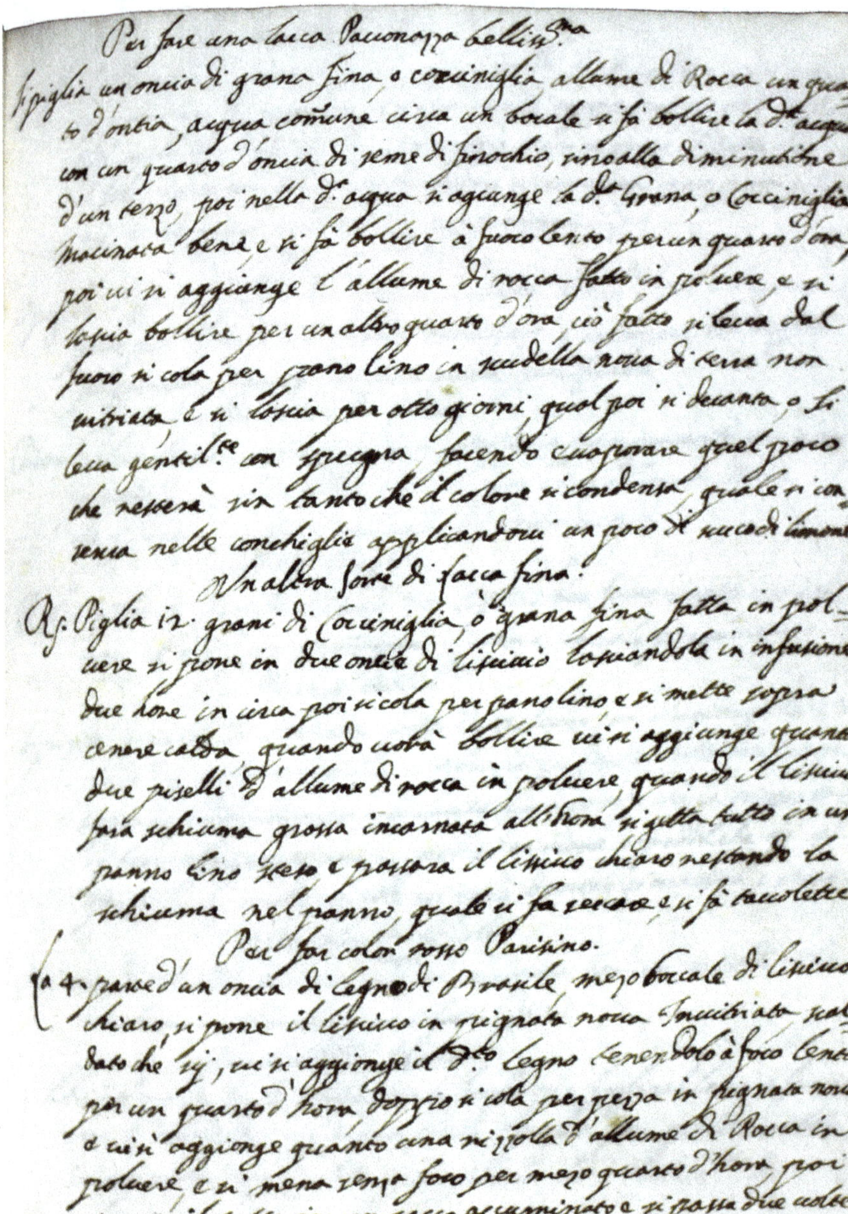

Per fare una lacca Pavonazza bellissima

Si piglia un oncia di grana fina, o cocciniglia, allume di Rocca un quarto d'oncia, acqua comune circa un boccale si fa bollire la d.ta acqua con un quarto d'oncia di seme di finocchio, sino alla diminutione d'un terzo, poi nella d.a acqua si agiunge la d.ta Grana, o Cocciniglia macinata bene, e si fà bollire à fuoco lento per un quarto d'ora, poi ui si aggiunge l'allume di rocca fatto in poluere, e si lascia bollire per un altro quarto d'ora, ciò fatto si leva dal fuoco si cola per panno lino in scudella noua di terra non uitriata, e si lascia per otto giorni, qual poi si decanta, o si leva gentil.te con spugna, facendo euaporare quel poco che resterà sin tanto che il colore si condensa, quale si conserva nelle conchiglie applicandoui un poco di succo di limone.

Un altra sorte di lacca fina.

R.: Piglia 12 grani di Cocciniglia, ò grana fina fatta in poluere si pone in due oncie di lissiuio lasciandola in infusione due hore in circa poi si cola per pano lino e si mette sopra cenere calda quando uorrà bollire ui si aggiunge quanto due piselli d'allume di rocca in poluere, quando il lissiuio farà schiuma grossa incarnata all'hora si getta tutto in un panno lino netto e posasi il lissiuio chiaro restando la schiuma nel panno, quale si fa secca, e si fà taccolette

Per far color rosso Parisino.

Ra 4. parte d'un oncia di legno di Brasile, mezo boccale di lissiuio chiaro, si pone il lissiuio in pignata noua inuitriata sul dato che sij, ui si aggionge il d.to legno tenendolo à foco lento per un quarto d'hora doppo si cola per pezza in pignata noua e ui si aggionge quanto una nizzola d'allume di Rocca in poluere e si mena senza foco per mezo quarto d'hora poi si pone il tutto in un sacco accuminato, e si passa due uolte come il uino coloro, e si conserva in conchiglie.

49

FIGURE 4.2 *Ms 992,* Another kind of fine lake, *fol. 49, c. 1600–1650, Padua, Biblioteca Universitaria*

which was copied from this manuscript also known and used by the painter (given that the author of the manuscript, in copying the text, decided to include also the dedicatory sonnet), and the second part of the manuscript, dedicated to lakes and varnishes, was probably copied from one or more seventeenth-century sources. Papers containing recipes of that kind were circulating among artists, and some of them might have been compiled in both the Merrifield manuscript and in Lomazzo's *Del Colore*.[45] The seventeenth-century manuscript, perhaps a collection of older and newer recipes, also includes different techniques, such as a whole section on varnishes for miniatures (fig. 4.3) and blue paper. The heterogeneity of the manuscript could be explained by concluding that this manuscript gathered recipes from different periods, which was not uncommon at the time. As Pamela Smith noted: "Recent scholarship has shown that practitioners wrote down their working methods and recipes for a variety of reasons aimed neither solely at providing utilitarian practical knowledge, nor necessarily with a view to teaching their trade to others."[46] These words certainly apply to the recipes for pigments presented above, although it can be argued that these collections were probably not used by artists in their workshops, as they are not detailed enough to be of practical use (for instance, they lack quantities and the procedures to follow in order to make those pigments). However, such recipes could have had a great variety of functions, besides teaching readers how to accomplish a specific task. In the case of Lomazzo's collection, we may speculate that he wished to give his readers an idea of the practical aspect of his work in order to legitimate his trade, in a form of self-advertisement and of a re-evaluation of the materiality of his activity as a painter.

The Transmission of Recipes for Colors in Sixteenth-Century Treatises on Art: A Short Overview

Lomazzo was not the only writer on art to include recipes for colors and technical suggestions on different painting techniques in his treatise. An increasing amount of literature on colors emerged in Italy during the sixteenth century,

45 For a recent discussion on the circulation of recipes and the transmission of artists' knowledge see Mark Clarke, Bert De Munck and Sven Dupré (eds.), *Transmission of Artists' Knowledge*.

46 Pamela Smith, "Sixteenth-Century Audiences for How-To: Who Read How-To Books?," *Reading How-To* workshop, Max Planck Institute for the History of Science, Berlin 19–20 September 2014.

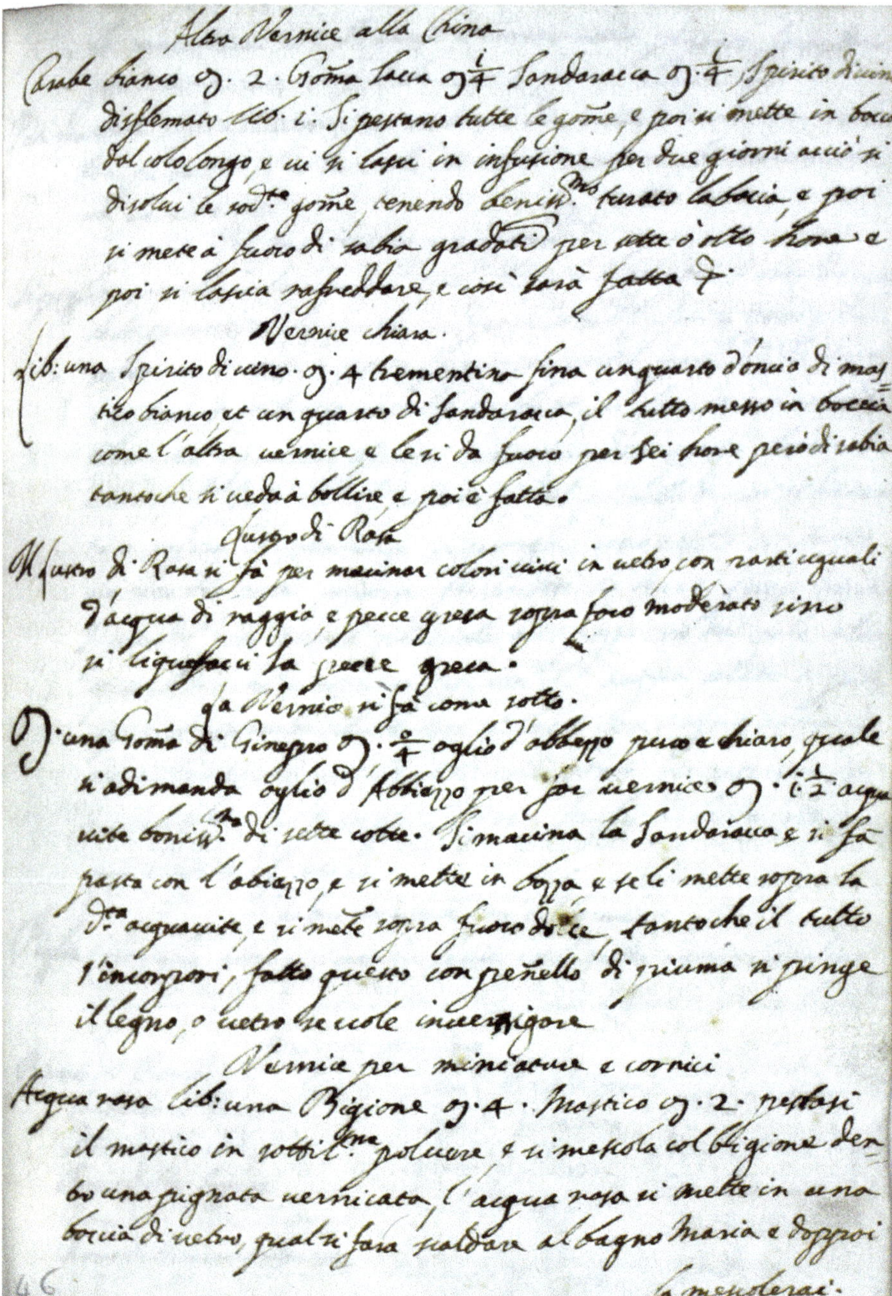

FIGURE 4.3 *Ms 992,* Varnish for miniatures, *fol. 46, c. 1600–1650, Padua, Biblioteca Universitaria*

mostly embedded in the tradition of treatises on art. These practical contents suggest an interest for this subject that goes beyond the workshops' walls. Practical suggestions on how to paint on different surfaces, as well as recipes for preparing and mixing pigments can be found in a relatively small but relevant part of the art literature of the period—usually in the form of dialogues or displayed in a few chapters within a more general theoretical discourse on the properties and the hierarchical place color should have among the parts of painting.

One of the first writers on art to comment on the difficulty of writing about practical aspects of coloring is Paolo Pino, who, in his *Dialogo di Pittura* (1548), states that:

> There are endless things that concern coloring and it is impossible to explain them with words, because every color, either mixed or not, can cause many different effects, and no color can imitate nature on its own, but only thanks to the intelligence and practice of the good master. And I wish to discuss with those who are experts in the arts, for this reason I will not discuss the kinds or properties of colors, since these are very clear to everybody, even those who sell pigments and who use them know all their qualities, and we have so many colors in every part of the world that talking about it would not be very helpful.[47]

Pino obviously does not add any suggestions on coloring or recipes, since he believed that "people (*meaning artists*) must know how to use both natural and artificial colors." Ultimately, from his words we gather that he does not seem to believe that writing down recipes solely for the practitioners' sake was of any use for his readers.

Nine years later, in his *Aretino* (1557), Lodovico Dolce makes a statement that places the value of coloring considerably higher than Pino:

---

47 Pino, *Dialogo*, p. 167: "Sono infinite le cose appertinenti al colorire et impossibile è isplicarle con parole, perchè ciascun colore o da sé o composito può far più effetti, e niun colore vale per sua proprietà a fare un minimo dell'effetti del naturale, però se gli conviene l'intelligenzia e pratica del buon maestro; et io, ch'intendo ragionare con chi è nell'arte perito, non m'istenderò altrimenti nella specie e proprietà de' colori, essendo cosa tanto chiara appresso ognuno, ch'insino quelli che li vendono sanno il modo di porli in opera e conoscono le qualità de tutti, sì minerali come artificiali, et anco n'è sì copiosa ciascuna parte dil mondo (oltre che Plinio et altri ne parlorono) che l'ispendervi parole non sarebbe molto profittevole."

> Coloring is needed for all those hues with which nature paints animate and inanimate things. (...) But I will discuss as a painter and not as a philosopher. (...) Those who possess this part of painting, have one of the most important parts. The main difficulty in coloring is to imitate the flesh which consists of a variety of hues (...) but no one should believe that the strength of coloring is choosing beautiful colors alone, such as good varnishes, blues, greens or similar.[48]

For Dolce, the choice of colors is not as important as the skill and expertise of the painter in mixing and combining them. In this book we do not find an extended discourse on the subject. Instead, it will be elaborated upon by the author eight years later, when he publishes a treatise entirely dedicated to color, the *Dialogo dei Colori,* in which we find an in-depth discussion on the theory of colors, but again not practical suggestions.[49] Pino of course gives some reasons why he thought his treatise was not the place to discuss the practice of coloring. Yet which reasons lay behind this choice?

The scholar and writer Giovanni Andrea Gilio gives a rhetorical answer to this lack of interest in technical aspects of colors, when he states, one year before Dolce's *Dialogue*, that he is skeptical about writing on the practice of coloring. This is mainly because, in contrast to ancient times, noble people no longer make art, and instead, it is now painters who try to make a living by it. Thus art may have lost some of the esteem that it used to have:

> Stated Silvio: that a nobleman goes to grind colors, to wear working clothes for the whole day, to have his hands oily and dirty, fearing that he will be censored and accused to do so for monetary gain: not considering that this noble art dignifies and distinguishes its practitioners.
>
> Of course this should not be, answered Vincenzo, because even Alexander the Great went to see grinding the colors by the pupils of Apelles. Polidoro added: when art was done by noble people this was not

48 Dolce, *Dialogo della Pittura,* p 167: "Il colorito serve a quelle tinte, con le quali la natura dipinge (che cosi si può dire) diversamente le cose animate et inanimate (...) Ma ragionerò da pittore e non da filosofo (...) Chi adunque ha questa parte, ne ha una delle più importanti. Così la principal difficultà del colorito è posta nella imitazion delle carni e consiste nella varietà delle tinte e nella morbidezza. (...) Né creda alcuno che la forza del colorito consista nella scelta de' bei colori, come belle lacche, bei azzurri, bei verdi e simili."

49 Dolce, *Dialogo dei Colori.*

> shameful, but now that others do it, it has not the credit among its followers anymore.[50]

The anecdote from antiquity should be considered only as a literary *topos*. However, this fictitious crossfire between the two characters conveys the message that the practice of "grinding colors" was an activity that was not highly regarded in Gilio's time. It would therefore not be a likely topic to include in a treatise that was concerned with demonstrating the nobility of the art of painting.

In order to find explicit practical suggestions on materials and techniques for painting, one has to turn to the more practical discourse by Leonardo, who in his *Trattato della Pittura*, lists some recipes and suggestions in twelve chapters of his book, such as how to paint with dry colors, prepare the paper, varnish, paint on canvas, re-use dried oil colors, prepare wood to paint on board, and recipes on how to prepare white and verdigris.[51] The material he presents is certainly interesting, though it is not very elaborate or systematized. Later, in in chapters XXIII and XXIV, towards the end of Biondo's Treatise *Della*

---

50 Giovanni Andrea Gilio, *Dialogo nel Quale si Ragiona degli Errori e degli Abusi de' Pittori* (Camerino: Antonio Gioioso, 1564), p. 6: "Rispose M. Silvio: 'Che un nobile si reca a mancamento il macinar colori, portar tutto il giorno il grembiale, aver le mani unte e lorde dagli olii; e di più teme di non essere di viltà tassato, ciò fare per guadagno: non considerando che questa nobilissima arte nobilita et esalta maravigliosamente gli suoi artefici.' 'Cotesto non dovrebbe essere, disse M. Vincenzo, perché Alessandro Magno (se vogliamo credere a Plinio) andava a vedere macinare i colori ai garzoni d'Apelle.' Soggiunse M. Pulidoro: 'Quando l'arte era in mano de' nobili, ciò fare non pareva vergogna; ma ora che gli par aver cangiato artefice, ha scemato anco il credito ai settatori suoi.'"

51 Leonardo da Vinci, *Trattato della Pittura* (ed. by Angelo Borzelli, first edition 1914), ch. XL–LII:

XLI	Per fare punte da colorire a secco
XLII	Il lapis si disfa in vino e in aceto o acquavite
XLIII	Carta da disegnare nero collo sputo
XLIV	Modo di ritrarre di rilievo
XLV	Per fare una pittura d'eterna vernice
XLVI	Modi di colorire in tela
XLVII	Per rinvenire i colori secchi a olio
XLVIII	Preparare il legname per dipingere su
XLIX	Bianco
LI	Del color verde fatto dalla ruggine de rame
LII	Aumentazione della bellezza nel verderame

Nobilissima Pittura (1549), we find recipes starting with a list of the surfaces for painting.[52]

One year later, the author of the *Vite dei Pittori*, Giorgio Vasari, inserts some technical suggestions in few chapters of his book, namely how to mix color when using oil technique, fresco and tempera, how to paint on walls, how to paint with oil on canvas, on dry wall and on board.[53]

Vasari presents in a discursive way the techniques of fresco, oil painting and how to achieve the "harmony of colors" following the principle of *decorum*. Other contemporary writers on colors, such as Mario Equicola, Fulvio Pellegrino Morato, or Coronato Occolti, limit the treatment of the subject to theory, mainly discussing (either in prose or in poetry) the origin and meaning of different colors.[54] One has to wait until the end of the century to find writers on art who included technical chapters in their theoretical writings: Cristoforo Sorte, of whom we have heard above, includes in his *Osservazioni nella Pittura* a discourse on which colors a painter should use to depict different subjects, a long letter dedicated to Bartolomeo Vitali, published in Venice in 1580.[55] In this letter Sorte starts a long discussion on the *materia dei colori*:

> But let us now turn to the material of colors. In order for our discourse to be easy to understand and logic, and for the things that we discuss to be

---

52 Biondo, *Della Nobilissima Pittura*, ch. XXIII, p. 20v: "Gli modi del pengere et sopra di che cosa al presente noi trattaremo, imperò lettor mio caro sappi che'l pittore ordisse la pittura quando sopra il muro sodo temperato, perciò con l'acqua, overo con la colla fatta di rettagli de carta peccorina, overo con quella fatta di rettagli di pelle di guanti, quando anchora stende sopra il muro secco con la tempera di l'ova, overo con l'oglio quando sopra il legno, et quando sopra la tela, lavora et penge con tempra di oglio et colla anchora, et questi sono gli modi e gli mezzi anchora del pittore nella pittura."

53 For the parts on color see Vasari, 1550, pp. 79–90; 1568, pp. 48–55, in Paola Barocchi, *Scritti D'Arte del Cinquecento*, 3 vols. (Milan and Naples: Ricciardi, 1971–1977), vol. 2, p. 2179. The chapters that present practical suggestions are the following:
XVIII – Come si debbino unire i colori a olio, a fresco o a tempera; e come le carni, i panni e tutto quello che si dipinge venga nell'opera a unire in modo che le figure non vengino divise et abbino rilievo e forza e mostrino l'opera chiara et aperta.
XIX – Del dipingere in muro, come si fa e perchè si chiama lavorare in fresco.
XX – Del dipingere a tempera overo a uovo su le tavole, o tele, e come si può usare sul muro che sia secco.
XXI – Del dipingere a olio in tavola e su tele.
XXII – Del dipingere a olio nel muro che sia secco.
XXIII – Del dipingere a olio su le tele.

54 See Barocchi, *Scritti*, pp. 2153–2198.

55 Cristoforo Sorte, *Osservazioni*, edited by Scuola Normale Superiore di Pisa (Pisa: Signum 2006).

> clear and useful, it is first important to know the ways in which you can use colors: those ways are four, that is *acquerello*, when you use colors on paper, *a guazzo*, when you paint on canvas, *a fresco* and *a secco*, that is on wall, and oil painting, that is when one paint on board, although sometimes one can also paint with this technique on wall.[56]

He then continues with a long list of recipes and suggestion on how to paint different subjects, such as waters, mountains and the sky.[57]

Beside Lomazzo, we find after Sorte the painter Giovanni Battista Armenini, who set down some recipes and practical suggestions in few chapters of his *Della Pratica della Pittura* (1586). These include ch. VII, on the distinction and kinds of colors, ch. VIII, how to prepare canvas and ch. IX, how to use oil painting.[58] However, none of these authors assign such a prominent place to their technical chapters as Lomazzo does, because he places them right after the chapters that discuss the definition of art and its status (on what art is, the virtue, and the necessity of each part of painting). Instead, the other authors usually placed the recipes and instructions for coloring at the end of their books.

Recipes and Technical Suggestions: the Colors of Paintings

As we have seen, Lomazzo copied most of these recipes and used them in the fourth chapter of *Del Colore* entitled "The materials in which colors can be found." These recipes describe the materials needed to make white, yellow, turquoise, green, *morello*, red, black, and "the shadow of the flesh", which usually consisted in different hues of pink, except for the depiction of dead bodies for which different hues of green were combined.[59]

---

56 Sorte, *Osservazioni*, pp. 281–282: "Ma passiamo alla materia dei colori, intorno alla quale, accioché i nostri ragionamenti siano facili et ordinati, e le cose, che ci occorreranno a trattare, meglio siano intese et alcuna utilità se ne possa indi cavare, fa di mestieri che prima sappiamo i modi generalmente di adoperar essi colori; li quali modi sono quatro, cioè: ad acquerele, il che aviene quando s'adoperano essi colori su la carta; a guazzo, quando si dipinge in tela; a fresco et a secco, nel muro; et a oglio, il che si fa ordinariamente quando si vuole operar su le tavole, benché anco si faccia a secco nel muro alcuna volta."

57 Sorte, *Osservazioni*, pp. 283–300.

58 Armenini, *De' Veri Precetti*, pp. 125–147.

59 Ciardi, *Scritti*, II, p. 168: "Quali siano le materie nelle quali si trovino i colori" (Lomazzo, *Trattato*, pp. 191–192).

The author states at the beginning that he will discuss the materials of which colors are made in his own time, and he presents a *rara invenzione*, that is the use of eggshell to make the fresco stay on the wall. This addition, which is not present in the Paduan manuscript, confirms the author's use of earlier sources in order to update the recipes, so that he could include some "workshop secrets":

> The matters of such colors as are generally used nowadays are, for the most part, known. And first of all the ones to make white are chalk, white lead and powdrered marble. And there is something else that cause the colors in Frisco to continue to stay put as if they were laid while the chalk is fresh: this is a rare secret (*invenzione*) which belongs to the practice of art, that is the white of an egg beaten very thin and mixed with all colors that shall serve to it.[60]

In this part I aim to make a comparison between the material Lomazzo presents with other recipes that can be found in sixteenth century writings on art, focusing on the terminology that the author uses and the recipes that he presents.

White

Lomazzo's ingredients for white are "chalk, white lead, white (probably meaning dry white) and powdered marble."[61] He does not include lime, pulverized eggshells or bone of the cuttlefish ground to a very fine powder, which are nevertheless included in the Paduan manuscript.[62] Lomazzo's list of materials is similar to the one in Michelangelo Biondo's *Della Nobilissima Pittura* (1549) when, in chapter XXIV, *Di varii colori,* he states that white is (made with the

60 Ciardi, *Scritti*, II, p. 168: "Tra i colori materiali che si usano generalmente a questi tempi se ne ha cognizione di molti, i quali tutti hanno i suoi particolari colori. E prima quelli che fanno il bianco sono il gesso, la biacca il bianco e il marmo trito. Evvi ancora un'altra cosa che a fresco fa restare i colori nel modo che si dipinge quando la calce è fresca, e questa è una delle rare invenzioni che sia nella pratica dell'arte, cioè il guscio delle uova tritato minutamente e con quello mescolare tutti i colori più e meno, secondo se gli appartengono." (Lomazzo, *Trattato*, p. 191).

61 Ciardi, *Scritti*, II p. 168.

62 Merrifield, *Original Treatises*, II, p. 648.

help of) white lead, chalk and dry white.[63] When describing *gesso*, an absorbent primer usually consisting of chalk, Lomazzo calls it a "friend of all colors, apart from verdigris, and so is the white lead, except for dry white."[64] This passage recalls Cennini's, which states that "white is a color that is made chemically (*or alchemically*), which is called white lead. [...] The the more you grind this color the more perfect and good on the panel it will be."[65] The discourse (as can be seen in appendix 8) is much more elaborate when compared to the recipes presented in the Paduan manuscript. By going into detail about the pigment in the technical part of his chapter on colors, Lomazzo wanted to include his expertise and know-how as a painter.

In contrast to the approach of many other art theoretical writers of his time to color, Lomazzo, in his discussion of white, never questions that white would be one of the seven principal colors. This seems to be an approach similar to Leonardo's who, more than half a century earlier, discussed whether white should be included as a color or not: "Black and white, although they are not put among the colors, because one is darkness and the other is light, that is one is deprivation and the other is generative, I cannot leave them behind, because in painting they are most important."[66]

Leonardo does not seem to have had a clear position on this matter, as later in his *Trattato della Pittura* he included white among the "simple colors": "The simple colors are six, and first is white, although some philosophers do not accept white or black as colors, because one is the cause of them and the other is the deprivation of them."[67] Other authors discussed this matter, which suggests that a discourse on what color was and which ones could legitimately be

---

63 Biondo, *Della Nobilissima Pittura*, ch. XXIV, *Di varii colori*, pp. 20r-21r.

64 Ciardi, *Scritti*, II, p. 169: "Giesso è amico di tutti i colori, eccetto che del verderame, la biacca similmente di tutti, ma è nemica del bianco secco." (Lomazzo, *Trattato*, p. 193).

65 See Cennini, *Libro*, p. 36: "Bianco è un colore archimiato di piombo, el quale si chiama biacca. (...) Questo colore quanto più il macini, tanto è piu perfetto, ed è buono in tavola."

66 Leonardo, *Trattato*, ch. XXI, *Della mistione delli colori l'uno con l'altro, la qual mistione s'astende in verso l'infinito*: "Nero, bianco, benchè questi non sono messi fra i colori, perchè l'uno è tenebre l'altro è luce, cioè l'uno è privazione e l'altro è generativo, io non li voglio per questo lasciare indietro, perchè in pittura sono li principali." See also ch. XXXII: "Perchè il bianco non è colore, ma è in potenzia ricettivo d'ogni colore." See Barocchi, *Scritti*, II, pp. 2125–2152. On the subject see also Alan Shapiro, "Artists' Colors and Newton's Colors," *Isis*, December 1994, *LXXXV*:600–630.

67 Leonardo, *Trattato*, ch. XXXV, *De colori che risultano dalla mistione d'altri colori, li quali si dimandano spezie seconda*: "I semplici colori sono sei, de' quali il primo è il bianco, benchè alcuni filosofi non accettino il bianco ne'l nero nel numero de' colori, perchè l'uno è causa de' colori e l'altro n'è privazione."

called natural colors was taking place in Italy at the time. Mario Equicola for instance, in his *Dal Libro di Natura d'Amore*, discusses what ancient philosophers said about colors, starting with the position of Democritus, who had already broached the question of whether white should be regarded as a color or not: "With Democritus then, or as Tullio (Cato) discusses, we think that the senses could really recognize colors, and of those that have no mixture there are of two kinds, black and white (if white is a color), the latter can easily be changed into another color, the former remains immutable."[68]

Leonardo had already noted the ability of white to take on parts of other colors, and this is probably why he called white generative. He points out this characteristic of white also when he describes how a white body should be depicted: "If you will paint a white body, this should be surrounded by much air, because white does not have color in itself, but is painted and transformed in that part of color that is there as object."[69] Here the painter denies that white is a color and states that it must be "surrounded by air" to become the color white you need, since by its nature it partially takes the color of the object in which it is embedded.

Lomazzo devoted another chapter to white towards the end of book III, which actually does not add much to the painterly use of the color, but links it with the concepts of purity, simplicity, and "as some say, height."[70] These concepts were not new in Lomazzo's time and they were already used in the fifteenth century. Earlier authors such as Dolce already drew similar symbolic analogies in their writings: "I will say now about white. This is the purest color, and it should be seen as a metaphor of the soul, which is sincere. This is the brightest color of all."[71] For these reasons, white was often used in the depiction of saints and holy figures. For instance, in Milan during that particular

68 Mario Equicola, *Dal Libro di Natura d'Amore*, 1526, pp. 162–165, in Barocchi, *Scritti*, III, p. 153: "Con Democrito adunque, benchè altrimenti in Tulio si dispute, crediamo li sensi conoscere veramente li colori, e di quelli senza mistione alcuna essere due spezie, bianco e negro (se'l bianco è color), l'uno facilmente in altro colore si muta, l'altro resta immutabile."

69 Leonardo, *Trattato*, ch. XII: "Se figurerai uno corpo bianco, (esso sia) circundato da molt' aria, perchè il bianco non ha da sé colore, ma si tignie e trasmuta in parte del colore che li è per obietto." In Barocchi, *Scritti*, p. 2131.

70 Lomazzo, *Trattato*, ch. XIII, p. 179.

71 Dolce, *Dialogo dei Colori*, p. 11: "Dirò hora del bianco. Questo è purissimo colore, la onde trasportandosi per via di metafora all'animo, si prende per sincero. Questo colore non si vede in altra cosa più chiaro, che nella neve."

time, different portraits of Carlo Borromeo were made dressing him either in white or red to underline his sanctity.

Black

As ingredients for black, Lomazzo lists "burnt ivory, the shells of burnt almonds, ball blacks, lamp blacks and black made of a kind of rubbish called black earth."[72] Compared to the Paduan manuscript, the ingredients are quite similar, apart from burnt ivory substituted with the "smoke of burned oil" (see appendix 8).[73] Biondo lists these ingredients as well: black earth, black from smoke and burnt bones of the perch fish.[74] The similarities between the manuscript and Michelangelo Biondo's chapter *Di vari colori* may suggest that he, too, almost thirty years before Lomazzo, might have taken information from the lost manuscript from which Lomazzo and the author of the Paduan manuscript copied different passages. In contrast to Lomazzo, Leonardo does not add recipes for black, but discusses which color is the right one to make a black shadow.[75] Again, Leonardo does not seem to consider black a color but the deprivation of all colors, and when applied to the canvas it creates, according to him, the strongest contrast with other colors.

Yellow

As materials for yellow, Lomazzo mentions the *giallolino di fornace di Fiandra et Alamagna* and the *oropimento*, which were both presented in Cennini's ch. XLVI *Della natura di un color giallo ch'è chiamato giallorino* and ch. XLVII, *Della natura di un giallo ch'è chiamato orpimento*, while Biondo only mentions the latter, along with "yellow earth and holy earth."[76] This part is substantially the same in the Paduan manuscript.

---

72 Ciardi, *Scritti*, II, p. 169: "Finalmente, quelli che fanno il nero sono l'avorio arso, il guscio della mandola, il nero di balla, il fumo di ragia, e finalmente il nero di scaglia detto terra nera."

73 Merrifield, *Original Treatises*, II, p. 651.

74 Biondo, *Della Nobilissima Pittura*, p. 20r.

75 See Leonardo, *Trattato*, ch. XI, *Qual colore farà ombra più nera*, in Barocchi, II, p. 2131.

76 See Cennini, *Libro*, p. 52: "Ogni colore di quelli che lavori in fresco, puoi anche lavorare in secco; ma in fresco sono colori che non si può lavorare, come orpimento, cinabro, azzurro della Magna, minio, biacca, verderame e lacca."

Green-Verdigris

Concerning the color green, Lomazzo uses the terms *verderame* or verdigris, verde azurro (light blue-green), a sort of light green (*verdetto*) that is called "holy yellow" and is similar to yellow, and green earth. He does not consider other materials for green that are copied in the Paduan manuscript, and are more related to botanical and medical traditions, such as *verde porro* (leek green), the *verde di vescica* (green of the bladder), the juice of the rue and of blue lilies. Again, Biondo used the same terms as Lomazzo in his description of green, but he also adds mountain green (*verdetto di montagna*). The colors *azzurro oltremarino, verderame and cinabrio* are already present in Cennini's discourse and are later mentioned by Leonardo as well.[77]

Light Blue – Turquoise

For the color turquoise, Lomazzo lists ultramarine blue, along with *smalti* from Flanders and *onagro*. This part is very similar to the manuscript and it is substantially the same in Biondo as well. The mention of glazing and of golden colors that are discussed in this chapter recalls an earlier tradition of making miniatures, which were out of date in Lomazzo's time, and were most probably not used in his workshop.[78] In Milan, the techniques of fresco, tempera and oil paintings were predominant both in the workshops of Milanese painters and of those of the "foreign painters", which were, as we have mentioned earlier in the book, in competition with the local painters for public commissions.[79]

Morello

The color *morello* is, according to the author, made from the *morello* of iron and salt, from burnt vitriol, from dark indigo and from *cilesto*, a pigment that I have not been able to identify. This latter ingredient is not mentioned in the manuscript, but instead *acquarella di tornasole* is listed. This pigment seems not to have been commonly known, since Haydocke in his translation of the

77 See Ciardi, *Scritti*, II, p. 169, n4, 5, 6.

78 See Merrifield, *Original Treatises*, II, p. 646. For an introduction on pigments in medieval recipes see among others Mark Clarke, *The Art of All Colors* (London: Archetype Publications, 2001), pp. 1–40.

79 See part I, chapter II.

Treatise (1598) felt the need to explain: "*Morello di ferro* and *di sale* doe makes a *morello* (which color is either bay or murrie) and so doth burnt vitriol, cilestro or sadazure, and darke indico."[80] Haydocke also adds a most interesting "footnote", which gives us an important clue not only to the nature of *morello* but on his method of translating the term, when he states that:

> My auctor seemeth here to put morello for a murrie or darke blewe color. Which me thinkes might be so understood, if he did not for the most part, ioyne *morello di ferro* and *morello di sale* together. For either the sinders of the smihes forge, or the scale which flie of in beating the red hot iron upon the anvil, being grounde, doe make a darke blewe color. But as far as *morello di sale*, it must neede be the rust of salt (…) There is a reddish color like unto rust digged out of the German salt mines, much desired of Painters, which peradventure is, ipse flos salis the flower itself of salt, for it is like it in color and raft, and is commonly called *morello di sale*. Wherefore I rather thinke it is the rust of iron, and the rust of salte, making naturally a bay color: for which cause I have still translated them the rust of iron and salte: though in some places they agree not in color as they are named in the mixture. So that I imagine there is some error crept into the booke, which by mine owne paines I cannot yet finde, nor by my conference with many good painters and chymisted.[81]

The appearance of *morello* is probably a sort of purplish-dark red, almost violet, and similar to the *pavonazzo*, as both terms are sometimes used as synonyms by other authors. For instance, Mario Equicola in his *Libro di Natura d'Amore* (1526), states that "purplish violet, that is *morello* or light *paonazzo*, makes rightfully the color of love, as it is the color of the favorite flower of Venus."[82] Also, Coronato Occolti in his *Trattato dei Colori* (1568), states that this color is frequently used in mixture with white, in order to "show passionate humility."[83] *Morello* seems to have been a color frequently used to create different shades of red, as we can see in the painting entitled 'Four Laughing Figures with a Cat' attributed to Lomazzo (fig. 4.4), and the painter thought that this color (*pavonazzo* or *morello*) could also stand for a "mad courage": in another

80 Haydocke, *A Tracte*, p. 99.

81 Haydocke, *A Tracte*, p. 100.

82 Mario Equicola, *Dal Libro di Natura D'Amore*, Venice 1526, in Barocchi, *Scritti*, II, p. 2157: "Violato purpureo, cioè morello o paonazzo chiaro, Amor li nostri meritamente fanno, per essere colore arridente del fiore grato a Venere dedotto."

83 Coronato Occolti, *Trattato de' Colori*, Parma 1568, in Barocchi, *Scritti*, II, p. 2205: "S'egli sarà accompagnato col morello, mostrerà appassionata umiltà."

part of book III the author states that it is made from "laque which has the color of the blood mixed with light blue", and that it also symbolizes the disdain of death for love. It would then be a sort of "ill red", which "through Mars shows animosity and madness."[84]

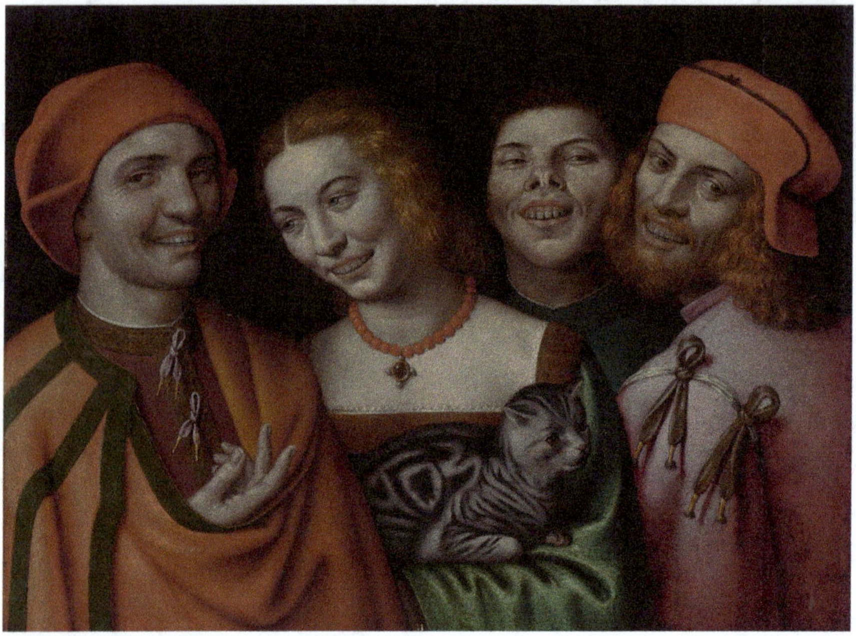

FIGURE 4.4 *Giovanni Paolo Lomazzo (attributed to),* Four laughing figures with a cat, *c. 1570, Sotheby's, New York (2012)*

Red

Lomazzo's discourse on reds is much more elaborate than the one in the Paduan manuscript. The latter only says "red is made with cinnabar, red earth and fine laque. *Ranzetto* is made with minio and burnt oropiment" (which is basically what Biondo suggests too). Lomazzo however, elaborates on the subject by adding new materials and definitions:

> Reds are made of the two cinnabars called vermilion, natural and artificial, and of the red earth called *Maiolica*. All lakes make sanguine red,

84 Ciardi, *Scritti*, II, p. 182: "Secondo alcuni denota dispregio di morte per amore, mostrando, come dicono, una certa pazza animosità, per la lacca color di sangue mischiato con colore azzurro." (Lomazzo, *Trattato*, p. 206).

> and the *ronzato* is made by red lead and orpiment, that is called golden color. And this is the alchemy of Venetian painters.[85]

This elaboration on the author's part is not uncommon, as we have seen in other parts of his writings. The links with the color red and alchemy should not be over-interpreted, as Lomazzo probably only wanted to underline the ability and skill (as such should the word *alchemy* be translated in this case) in coloring and, in particular in making and using red. Notably, Lomazzo had seen the works of Venetian painters, first of all of Titian.

His words are similar to Cennini's when he explains the red color *cinabrio*:

> Red is a color that is called cinnabar. And this color is made through alchemy, with the alembic. Of this, since it would be too long to explain every way and recipe to do it, I will not. Why? Because if you want to trouble yourself, you can find many recipes for it especially among the friars.[86]

Shadow of the Flesh

Finally, for the materials to make the *umbra delle carni* (fig. 4.5), Lomazzo borrows from the manuscript umber earth, aspaltum and green burnt earth to form his definition of the former (*detta falzalo*), and adds the popular ingredient of the mummy. The author returns to the discourse on carnation and its shadow when he discusses the practice of painting, ch. X, entitled "How shadows should follow the color of the flesh." In this part, the author states that: "The shadow of the flesh is nothing but flesh that is not enlightened" and he links the four Aristotelian humors to the four mixtures that apply to the color of flesh, making the following connections:

85 Ciardi *Scritti*, II, p. 169: "Quelli che fanno il rosso sono i due cinabri, cioè di minera et artificiale, e la terra rossa, detta maiolica. Il rosso sanguino lo fanno le lacche tutte, et il ronzato lo fa il minio et ancora l'orpimento arso, il quale si dice color d'oro. E questa è l'alchimia dei pittori veneziani." (Lomazzo, *Trattato*, p. 191).

86 Cennini, *Libro*, p. 26: 'Rosso è un colore che si chiama cinabro: e questo colore si fa per archimia, lavorato per lambicco; del quale, perchè sarebbe troppo lungo a porre nel mio dire ogni modo e ricetta, lascio stare. La ragione? Perchè se ti vorrai affaticare, ne troverai assai ricette e spezialmente pigliando amista di frati.' In Barocchi, *Scritti*, II, 2149, n3.

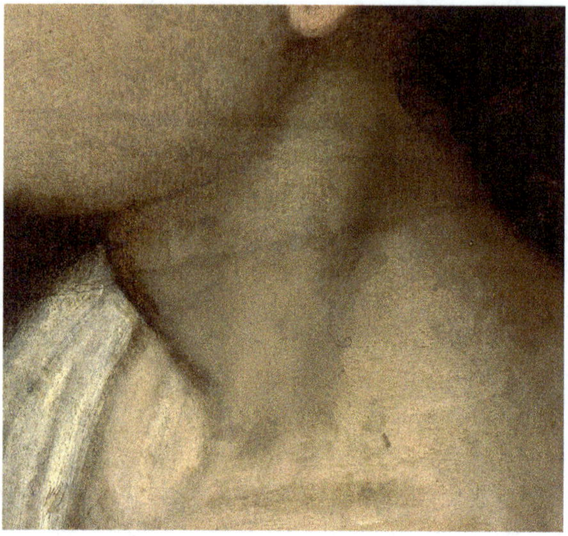

FIGURE 4.5
Giovanni Paolo Lomazzo, Self portrait as a young boy, *detail, c. 1570, Vienna, Kunsthistorisches Museum*

> Melancholic carnation -umber earth
> Phlegmatic carnation -white with (occasionally) green or light blue
> Sanguine carnation -white and red, which will result in pink
> Choleric carnation -extreme red, such as cinnabar or lake, adding white[87]

The Techniques for Painting

Lomazzo links all the *colori* presented above to different techniques of painting, and he states that it is important to know which sorts of pigment can be used with which binding medium:

> Because some colors cannot be used in all three kinds of paintings, as in Frisco, which is done upon fresh chalk, oil, and tempera, otherwise they will die, I will distinguish them, as they agree with each of them.[88]

87 Ciardi, *Scritti*, II, p. 270, *Come l'ombre debbono seguire il colore delle carni*: "Dico che quei quattro colori che rappresentano i quattro umori e le quattro qualità degli elementi soprannominati con le misture loro, per far le carni melanconiche sono come le terre d'ombra e simili; per le flemmatiche è il bianco, che s'accompagna secondo le occorrenze col verde et azzurro; per le sanguigne la mischia fatta di bianco e rosso che risulta in color rosato; e per le coleriche il rosso estremo, come la lacca e il cinabro; ma in modo che, spargendosi con molto bianco, ne riesca un color pallido che imiti il colore della fiamma spenta." (Lomazzo, *Trattato*, pp. 311–312).

88 Ciardi, *Scritti*, II, p. 169: "Perchè alcuni colori non si possono adoperare senza la morte loro in tutte e tre le spezie di dipingere, che sono il fresco sopra la calce fresca, il lavorar a oglio

Lomazzo again makes the distinction between natural colors and artificial colors. As for "the artificial practice of colors", he does not explain it thoroughly, since, as he states, "Bernardino Campi wrote a large and learned treatise, whose industrious work demonstrates his great knowledge in the true practice thereof."[89] It is significant that Lomazzo decided to mention Bernardino Campi, a painter from Cremona, who was therefore part of the "foreign competition", and had different public commissions in Milan. Lomazzo wanted to surpass the work of previous writers on art with his book, and the fact that Campi wrote a treatise dedicated to colors (unfortunately now lost) testifies to painters' desire to speak and write about this subject beyond the workshop's walls, probably in order to surpass the literary achievements of foreign painters.

In chapters VII and VIII, *Quali colori e meschie faccino l'un colore con l'altro* and *Della convenienza c'hanno fra loro i colori chiari et oscuri*, the author again presents a list of recipes possibly taken from the source that was also used by the author of the Paduan manuscript. If we compare the two, we shall see that Lomazzo adds at the beginning of the chapter a paragraph discussing how to work with oil which is not present in the manuscript: "My meaning is not to speak of the mixtures of colors as they concern each kind of work particularly, but only of those which belong to oil. Because from these you may draw the observation for other sorts of works; by mixing such colors which agree with that manner of work you have in hand."[90] In this passage we hear Lomazzo's direct voice, as we do when he includes some clarifications, while copying the recipes to make white lead. The Paduan manuscript thus describes the preparation of white lead:

> The composition and mixtures of the colors, and first of white lead. *Sbiadato* is made with white lead mixed with ochre, and is a color similar

et il lavorar a tempera, gl'anderò distinguendo secondo che a ciascuna di queste tre maniere di dipingere si convengono e si comportano." (Lomazzo, *Trattato*, p. 192).

89 Ciardi, *Scritti*, II, p. 170: "Ma del porre in opera con diligenza et arte i colori per ciascuna sorte di lavorare, Bernardino da Campo, cremonese, ne ha fatto un copioso e diligente trattato e lo ha saputo anco mettere in pratica nelle opere sue, fatte con cura grandissima." (Lomazzo, *Trattato*, p. 193).

90 Ciardi, *Scritti*, II, p. 172: "Ma solamente ragionerò di quelle che si appartengono al lavorar ad oglio; dal che si potrà poi cavarne regola per ogn'altra spezie di lavorare, componendo sempre i colori del medesimo colore conforme alla spezie del lavorare." (Lomazzo, *Trattato*, p. 194).

> to straw, or white lead with terra gialla. *Biondo* is made with white, ochre and *gialdolino*.[91]

Lomazzo uses the following words to explain how to make white lead:

> First white lead mixed with yellow ocker, makes a pale straw color, and by adding a little yellow, it makes the lights of the blonde color, much like boxe.[92]

He includes his painterly expertise in the text, and adds information to many of the recipes that he borrows from his literary source. Surely, as Ciardi stated, some subjects that the painter discussed in his book on color were also popular before the author's time, such as the discourse on the *colori cangianti* and their use (ch. X), which again can already be found in Cennini.[93]

As we have mentioned above, the techniques that were most commonly used for private commissions in Milan at the time were either oil on canvas or tempera on panel, whereas for public commissions the majority of painters worked with frescoes, such as those executed by Lomazzo in Santa Maria Presso San Celso around 1568 or in the Foppa chapel in the church of San Marco in 1570 (see fig. 5.6).[94] This constitutes a clear hint of the author's tendency to collect and assemble literary material from previous periods, which were probably not meant to be of directly useful for artistic practice, but should be nonetheless known for the sake of the painter's education.[95]

On the other hand, there are some useful practical suggestions for painters, for instance about transparent colors (ch. IX). Lomazzo recommends the use of varnishes and verdigris when working with oil, and states that these are wonderful "in order to represent most natural lights and shadows." He advises

---

91 Merrifield, *Original Treatises*, II, p. 650–651: "Compositione e mischia di colore e poi della biaca. Sbiadato si fa con biaca mischiata con l'ocrea, et è color simile alla paglia, o biaca con terra gialla. Biondo si fa con biaca, ocrea, e gialdolino."

92 Ciardi, *Scritti*, II, p. 171: "E per cominciare, si truova che la biaca, mischiata con la ocrea, fa color sbiadato, l quale e simile alla paglia, et aiutato dal giallolino, fa la luce del color biondo e simile al busso." (Lomazzo, *Trattato*, p. 194).

93 See Ciardi, *Scritti*, I, p. 175, n1.

94 The issue of Lomazzo's pictorial production, as stated in the first part, deserves a research in itself. I refer here to the latest study that I know of on the subject, that is the already cited doctoral thesis *Giovan Paolo Lomazzo Pittore Milanese* by Dr. Mauro Pavesi.

95 As we have already seen in the first part, Lomazzo has an idea of "complete education" for the artists, which includes both technical training and literary education.

his readers to use *verderame* or *verdetto* to represent transparents bodies such as emeralds.[96]

Chapter XI, *De gl'effetti che causano i colori*, discusses the effects of colors on temperaments, linking black with melancholy, red with courage and yellow with attentiveness, a discourse which will be, as the author concludes "further developed in the book of practice."[97]

From chapter XII to the end of the book, Lomazzo describes the principal colors as they were used in antiquity, and brings up the issue of *decorum*. He begins with black—"Athenians used to wear black when some misfortune happened" and white – "Virgil writes that chaste priests are dressed in white, such as the worthy poets", and does not add any practical advice.[98] For this part Lomazzo, as Ciardi already noted, takes most of his information from *Del Significato dei Colori*, a treatise by the Mantuan writer Fulvio Pellegrino Morato.[99]

Colors, Planets and Elements: the Astrological Connections

In the analysis presented so far, I have deliberately omitted the astrological connections that Lomazzo makes between planets, elements, temperaments and colors, and for which he is best known. The reason for this is that the author in fact does not draw these connections in *Del Colore*. Apart from chapter XI, which we have discussed above, he hints at planets once, in his description of the color *morello*:

> The color *morello* or *pavonazzo*, which means what I have already stated elsewhere, denotes death which despises love, showing, as some say, a

96 Ciardi, *Scritti*, II, p. 173: "Con la medesima via ancora il verderame et il verdetto avvivano e rappresentano la temperanza de gli smeraldi e simili materie trasparenti. I medesimi colori si usano ancora per dar il lustro e la vivacità al raso et all'ormesino alterati dei loro colori naturali sopra le abbozzature." (Lomazzo, *Trattato*, p. 197).

97 Ciardi, *Scritti*, II, p. 177: "Ma come vadano compartiti nelle istorie e per tutte le opere si dirà, più brevemente e chiaro che si potrà, nel libro della prattica." (Lomazzo, *Trattato*, p. 202).

98 Ciardi, *Scritti*, II, pp. 178–179: "Solevano gl'Ateniesi, quando accadeva loro qualche sciagura, vestirsi di color oscuro (…) Di lui, scrivendo Virgilio nel sesto, ne veste i sacerdoti casti, i buoni poeti e gl'uomini d'ingegno e della patria difensori." (Lomazzo, *Trattato*, pp. 202–203).

99 Fulvio Pellegrino Morato, *Del Significato dei Colori* (Venice: Giovanni Antonio Nicolini da Sabbio, 1535).

> certain mad animosity, as the blooded-colored laque is mixed with light blue, so that the result is a mixed color, in between Jupiter and Saturn, the former showing animosity and madness through Mars.[100]

While Lomazzo only mentions the planets Jupiter, Saturn and Mars in connection with a particular color, he explains the complete connections of the planets, the elements and their respective colors later in the Treatise, in the book *Della pratica della pittura.*[101]

In the first few chapters of this book, which, according to the artist was meant to be a more practical handbook for artists, the connections between colors, planets and elements are explained as follows: black is connected with Saturn, melancholy and earth, Jupiter with blue (rather ultramarine light blue), glory and air, Mercury with *purpureo* and continence, the Sun both with yellow and gold as well as with light red and courage, dark red is linked to Mars and fury, Venus to green, hope and water, and the Moon is connected with white, silver, phlegm and again water (fig. 4.6):

> The first color is yellow, it is dedicated to the sun, since it is similar to its ray and to gold, the first of metals, as we know, and the most precious. And sun means nobility, richness, religion, clarity, *gravità*, justice, faith and corruption. White means and represents innocence, purity, and in the man the phlegm, in the seasons the autumn. Among the virtues, since it is the immaculate color, it means justice; among the elements it represents water, and among metals silver, and among theological virtues hope, that has to be pure and neat.
>
> Red, which represents among the elements fire and among the planets sun, means courage, height, victory, blood, martyrdom, when we lean towards the darker red of Mars; in the man it means fury, among the theological virtues charity, and among the seasons summer. Light blue, corresponding to Jupiter, means the overall complexity of the blood system, height, glory, dignity, sincerity, happiness and so on. Among the elements it is linked to air. Black means melancholy, sadness, grief,

100 Ciardi, *Scritti*, II, p. 182: "Il color morello o pavonazzo, che veramente significa quello che altrove si è detto, secondo alcuni denota dispregio di morte per amore, mostrando, come dicono, una certa pazza animosità, per la lacca color di sangue mischiato con color azurro, onde si compone l'uno colore mezzano tra Giove e Saturno, il primo dei quali per Marte mostra l'animosità e la pazzia." (Lomazzo, *Trattato*, p. 206).

101 Ciardi, *Scritti*, II, pp. 401–402 (Lomazzo, *Trattato*, book VI, chapter LVIII, *Compositione dei Colori*).

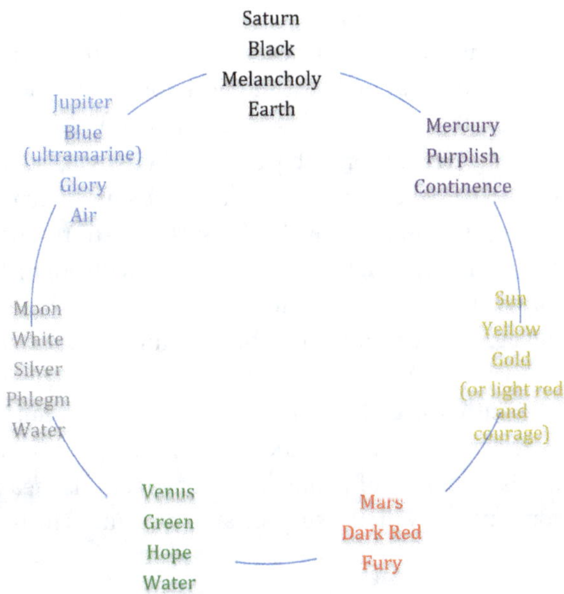

FIGURE 4.6
Lomazzo's system of colors

stability, its planet is Saturn and its season winter. Among the virtues it is prudence and among the elements earth, and that is also yellow because of the dryness. (...) Green is spring, Venus, gaiety, *vaghezza*, hope, goodness, jocundity and similar, among the ages it is youth, water is its element. *Porpora*, which is composed by all the aforementioned ones, and that is that color we call dry rose, such as Sicilo Araldo states, which is linked to Mercury and means, since it contains all others, triumph, honor, principality and similar.[102]

102 Ciardi, *Scritti*, II, pp. 401–402: "Il primo colore adunque è il giallo, dedicato al sole, per assomigliarsi ai suoi raggi et all'oro, principal metallo, come si sa, di tutti è più grave. E perciochè il sale, se ben nel suo centro e più tinto di rosso, ha però i raggi che ritirano più al secco della terra, significa nobiltà, ricchezza, religione, chiarezza, gravità, giustizia, fede e corrozione. Il bianco significa e rappresenta innocenza, purità, e nell'uomo si dipinge per la flemma, nelle stagioni per l'autunno; fra le virtù, per essere colore immaculato, significa anco la giustizia; fra gl'elementi rappresenta l'acqua, e fra i metalli l'argento, e fra le virtù teologiche la speranza, che deve esser pura e netta. Il rosso, che fra gl'elementi rappresenta il fuoco e fra i pianeti il sole, significa ardire, altezza, vittoria, sangue, martirio, maggiormente inchinando al rosso più oscuro e fosco di Marte; nell'uomo mostra la colera, nelle virtù teologiche la carità, che deve essere accesa d'amore et ardente e fra le stagioni rappresenta l'estate. L'azurro oltramarino, che risponde a Giove significa la complessione sanguigna, dimostra altezza, gloria, dignita, sincerità, allegrezza e simili; e

Ciardi states that this part is taken from Morato's *Del Significato dei Colori.* Lomazzo certainly had direct knowledge of Morato's work, since the former used an exact quotation from Morato's sonnet at the beginning of his book, stating that: "Some modern writers say that this color (*morello*) means to despise death in favour of love."[103] He was probably inspired by Morato's titles of the different colors, as he also linked colors to temperaments, some with the same words (such as the example of Morello, which "despises death through love"), and others with different ones (linking for instance yellow with hope).[104]

The similarities with Morato are mostly formal, though; Lomazzo's system is therefore original to a great degree. The sources used by the painter are considerably more diverse than those that Morato used. We may have an idea of Lomazzo's sources when the author lists the authorities for his discussion of colors, among which are Platonics, Aristotelians, Lucretius, Pliny, Equicola and again Morato.[105] However, this catalogue of names, as we have seen, is meant to be evidence of the author's authority on the subject, and the sources he had direct access to may well go beyond those that are listed here.

---

ne gl'elementi l'aereo. Il nero significa melancolia, tristezza, duolo, gravità e stabilità, et il suo nume è Saturno; e delle stagioni rappresenta il verno, delle compassioni la melancolia, delle virtù la prudenza, de gl'elementi la terra, che ancora si mostra col giallo per la sua siccità; delle età la decrepita, e de gl'accidenti la morte che significa divisione e separazione. E volendo scrivere, o disegnare, col colore oscuro si va partendo la carta per quegli spazi che si fanno. Il verde, che dimostra la primavera, e risponde a Venere, significa allegrezza, vaghezza, speranza, bontà, giocondità e simili, nelle età la gioventù, e de gl'elementi è dato parimenti all'acqua. La porpora, colore composto di tutti i sopradetti, e che non è altro che quel colore che chiamiamo rosa secca, come dice Sicilo Araldo, è data a Mercurio e significa, per contenere tutti gl'altri, trionfo, pregio, onore, principalità, e simili." (Lomazzo, *Trattato*, pp. 464–465). For a similar scheme of Lomazzo's colors see Kemp, *The Science of Art*, p. 272, although the colors that he lists are partly different from the list I give here.

103 Ciardi, *Scritti*, II, p. 402: "Il morello morte per amor disprezza." (*Trattato*, 1584, p. 465) and Morato, *Del Significato*, p. 2.

104 Morato, *Del Significato*, index of colors:
"Il color Verde esser ridutto a niente dimostra
Il Rosso ha poca sicurezza
Il Nero ha il suo voler pien di matezza
Il Bianco ha suo appetito et voglie spente
Il Giallo ha sua speranza rinascente
Il Morel Morte per Amor disprezza
Il Torchino ha'l pensier molto elevato."

105 Ciardi, *Scritti*, II, p. 401: "I platonici, gli aristotelici, Lucrezio, Donato, Marco della frata, Plinio, Mario Equicola, Virgilio, Servio, Telesia, Marcello, Il Falcone, Fulvio Morato, Arrigo et altri." (Lomazzo, *Trattato*, p. 463).

The connections between colors and elements were a topic in the well-known treatise *De Coloribus*, which was thought to be by Aristotle:

> Those colors are simple which belong to the elements, fire, air, water and earth. For air and water are naturally white in themselves, while fire and the sun are golden. The earth is also naturally white, but seems colored because it is dyed. This becomes clear when we consider ashes; for they become white when the moisture which caused their dyeing is burned out of them; but not completely so, for they are also dyed by smoke, which is black. In the same way sand becomes golden, because the fiery red and black tints the water.[106]

It is possible that the painter had access to the original text, or that he had knowledge of its content from a mediated source. Whatever the case, Lomazzo is the first writer on art to build such a complex system of analogies between colors, planets, temperaments and elements. In building this system he was also probably inspired by the work of Cornelius Agrippa von Nettesheim, and especially by the famous book I of *De Occulta Philosophia* entitled *Magia Naturalis* (1531), in which Agrippa explains the theory of the four elements:

> There are four Elements, and original grounds of all corporeal things: Fire, Earth, Water, Air, of which all elementated inferiour bodies are compounded; not by way of heaping them up together, but by transmutation, and union; and when they are destroyed, they are resolved into Elements. For there is none of the sensible Elements that is pure, but they are more or less mixed, and apt to be changed one into the other: even as Earth becomes dirty, and dissolves, it becomes Water, and the same is made thick and hard, and becomes Earth again; but being evaporated through heat, passes into Air, and that being kindled, passes into Fire, and this being extinguished, returns back again into Air, but being cooled again after its burning, becomes again Earth and Stone and Sulfur.[107]

106 Theophrastus (Pseudo-Aristotle), *De Coloribus*, 793b34–794a15 in Aristotle, *Minor Works*, translated by W.S. Hett, Loeb Classical Library (Cambridge (Mass.): Harvard University Press, 1952). On the issue of autorship, see H.B. Gottschalk, "The *De Coloribus* and its Author," *Hermes*, 1964, 92:59–85. For a recent discussion on this text see Karin Leonhard, *Bildfelder: Stilleben und Naturstücke des 17 Jahrhunderts* (Berlin: Akademie Verlag, 2013), pp. 340 ff.

107 Vittoria Perrone Compagni (ed. by), *De Occulta Philosophia Libri Tres* (Leiden, Brill 1992), book I, ch. III, p. 89: "Quatuor sunt elementa et primaria fundamenta rerum omnium corporalium: ignis, terra, aqua, aer, ex quibus omnia elementata in istis inferioribus

The elements are linked to planets and temperaments later in the book (in order to understand which parts of the body correspond to a certain planet):

> Also all things under Saturn conduce to sadness, and melancholy; those under Jupiter to mirth, and honor; those under Mars to boldness, contention, and anger; those under the Sun to glory, victory and courage; those under Venus to love, lust, and concupiscence; those under Mercury to Eloquence; those under the Moon to a common life. Also all the actions, and dispositions of men are distributed according to the Planets.[108]

These connections must have inspired Lomazzo, who was an admirer and reader of authors such as Agrippa, Cardano and Giovan Battista della Porta. As Kemp puts it: "Much of Lomazzo's celestial magic may seem to have nothing to do with science as we understand it, but a system such as his suffers grave distortion if we attempt to separate his 'science' from his 'magic' in an anachronistic manner."[109]

Although Lomazzo's systematization of colors and their connections, as we have seen, yield quite original results, if one takes into account only the connections between colors and elements it is easy to see that the author's discourse derives (at least partly) from connections between colors and elements by earlier writers on art, as the following table shows (fig. 4.7).[110]

Despite the importance that later critics attached to the cosmological system in Lomazzo's writings and theory, these connections are only implied in *Del Colore* and never explicitly stated. Lomazzo might have purposely avoided

componuntur, non per modum coacervationis, sec secundum transmutationem et unionem; rursusque, cum corrumpuntur, in elementa resolvuntur. Nullum autem sensibilium elementorum purum est, sec secundum magis et minus permixta sunt et in se invicem transmutabilia, quemadmodum terra lutescens dissolutaque fit aqua atque illa ingrossata densataque fit terra, per calorem autem evaporata transit in aerem et ille supercalescens in ignem et hic extinctus revertitur in aerem, infrigidatus autem ex superadustione fit terra aut lapis aut sulphur."

108 Cornelius Agrippa von Nettesheim, *De Occulta Philosophia* (ed. by Vittoria Perrone Compagni), I, ch. XXII, p. 130: "Praeterea conferunt saturnalia ad tristitiam et melancholiam, iovialia ad laetitiam et dignitatem, martialia ad audaciam, rixam et iram; solaria ad gloriam et victoriam et animositatem; venerea ad amorem, libidinem et concupiscentiam; mercurialia ad facundiam; lunaria ad vulgarem vitam. Ipsaque hominum exercitia et mores secundum planetas distributa sunt."

109 Kemp, *The Science of Art*, p. 270.

110 This scheme is partly taken from the scheme by Götz Hoeppe in his book *Why the Sky is Blue: Discovering the Color of Life* (Princeton: Princeton University Press, 2007), p. 57.

explaining his complex system of symbolic associations in this book in order to keep his discourse on the subject more systematic, in the same way that he organized his discussions for movement, light, perspective and proportion.

Author	Earth	Water	Air	Fire
Leonardo 1510	Yellow	Green	Light Blue	Red
Equicola 1526	Green	Purple	Yellow	Red
Dolce 1564	Dark/Black	Transparent	Transparent	Yellow
Lomazzo 1584	Black	White + Green	Light Blue	Red

FIGURE 4.7 *Examples of systems of colors and elements in sixteenth-century Italian writings on art*

Conclusions

Lomazzo in *Del Colore* integrates artistic terminology with different bodies of knowledge in order to systematize the information he possessed on the subject. At the same time, the author decided to include recipes from an earlier literary source, although these were probably no longer used in the practice of the *bottega*. In this way, these recipes became part of his discourse on the theory of color. As Barasch already noted, despite its declared theoretical setting, Lomazzo's thought does not completely distance itself from the reality of the workshop.[111] He uses pictorial works that he is familiar with in order to exemplify the seven principal colors that a painter must be able to obtain and use. His practical suggestions on the manners of coloring are found in what has usually been labeled as his more "philosophical" treatise, the *Idea del Tempio della Pittura*, whereas astrological connections are listed in the book that was intended to be more useful for painters, that is *Della Pratica della Pittura*. One explanation for Lomazzo's inclusion of more theoretical aspects of color may be that he perceived a connection of colors, planets and elements as a substantial part of the *pratica* of artists that for him should go beyond the knowledge

111 See Barasch, *Light and Color*, p. 158 (and p. 139 ff. for the analysis of light).

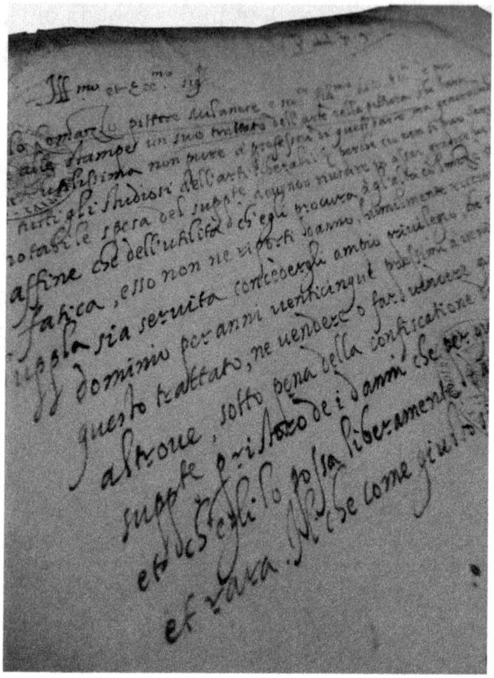

FIGURE 4.8
Giovanni Paolo Lomazzo, Request of Privilegio, *1582, Milan, Archivio di Stato*

of how to mix colors and how to obtain the right tone for the different parts of a painting. Since the author borrowed concepts of colors from a variety of different sources, ranging from literary and oral sources, he offered his readers new theoretical facets and at the same time practical aspects of color. He drew mainly on collections of recipes in manuscript form, terms and suggestions from the reality of the workshop, previous philosophical writings and previous writings on art. He achieved an idea of color that integrated artistic experience and practical recipes in discursive knowledge, which would have been part of the discussions that artists and men of letters were leading regarding the art of painting in Lombardy.[112] The *mélange* of the theoretical discourse on colors with technical recipes should perhaps be interpreted, on one hand, as a result of the painter's general attempt to upgrade his status as a writer on art by way of writing a book on the "science of painting", which painters allegedly should read to be included in the author's ideal of the "good artist." On the other hand, it might have been written in order to be of service to the *professori del disegno* and for the lovers of the liberal arts to obtain a proper vocabulary to speak

112 On this see James Ackerman, "On Early Renaissance Color Theory and Practice," in *Distance Points: Essays in Theory and Renaissance Art and Architecture* (Cambridge (MA): The MIT Press, 1991).

about painting, as to these *professori* the painter addresses in the request of *privilegio* for his treatise, dated 1582 (fig. 4.8).[113] Whatever the case, we have seen how the topic "color" cannot be confined to the book *Del Colore* but it is closely connected with other parts of his writings and especially with light. This is the case also for perspective, as we shall see in the next chapter.

---

113 ASM, marti 9 october 1582, in Bora, *Rabisch*, p. 333: "Ill.mo et ecc. mo sign. Gio Paolo Lomazzo pittore Milanese (...) è per dare alle stampe un suo trattato dell'arte della pittura che sarà opera utilissima non pure a' professori di quest'arte, ma generalmente a tutti gli studiosi dell'arti liberali."

CHAPTER 5

Acutissima è La Prospettiva

Lomazzo's Discourse on Perspective and his Connections with Contemporary Practitioners and Men of Science

In 1572, the Milanese architect Martino Bassi published a treatise on perspective entitled *Dispareri in Materia d'Architettura e Prospettiva con Pareri di Eccellenti e Famosi Architetti,* which consists of a fictitious exchange of letters between him and an unknown lord from Verona, "Messer Alfonso N". This interesting text is the result of a quarrel that took place in Milan three years before, in 1569, on the perspective to be adopted for the relief of the Annunciation, which was to be placed on one of the *timpani* in the *Duomo*. It is probably the most renowned episode that led to a discussion on the "proper perspective" regarding a religious ornament in the city at the time. The debate saw Bassi in disagreement (and in competition for the job) with another famous architect, Pellegrino Tibaldi.[1] The latter was called from Ancona and appointed architect of the cathedral by Carlo Borromeo in 1567, and he wanted to place the relief at the very center of the tympanum, modifying the previous perspective in favor of a closer distance with the spectator. On the other side was the Milanese architect, Bassi, who was in favor of an illusionistic scheme to be viewed from below. As a possible alternative, the latter also suggested a "central view" with a central vanishing point. In his book Bassi accused the foreign Tibaldi of lacking "rule" (*regola*) in his scheme and of operating "with-

1 See Martino Bassi, *Dispareri in Materia d'Architettura e Prospettiva con Pareri di Eccellenti e Famosi Architetti* (Brescia: Marchetti, 1572). This episode is mentioned in Filippo Picinelli, *Ateneo dei Letterati Milanesi* (Milan: Francesco Vigone, 1670), p. 414, and later in Filippo Baldinucci, *Notizie de' Professori del Disegno* (Florence: D.M. Nanni, 1768–1820), 6 vols., vol. 6, p. 415. On this debate see Erwin Panofsky, *Die Perspektive als Symbolische Form* (Leipzig-Berlin: Vorträge der Bibliothek Warburg", Leipzig-Berlin, 1927), pp. 108–110, Christopher S. Wood (New York: Zone Books, 1991); Bora, "La Prospettiva della Figura umana: gli 'Scurti' nella Teoria e nella Practica Pittorica Lombarda del Cinquecento," in Marisa Dalai Emiliani (ed. by), *La Prospettiva Rinascimentale: Codificazioni e Trasgressioni* (Florence: Centro Di, 1980), 1: 295–317; Kemp, *The Science of Art*, pp. 73–74, and more recently Pascal Dubourg-Glatigny, "La Question de la Perspective Matérielle dans les Traités du Cinquecento," in Marianne Cojannot Le Blanc, Marisa Dalai Emiliani & Pascal Dubourg-Glatigny (eds.), *L'Artiste et l'Oeuvre à l'Epreuve de la Perspective* (Rome: École française de Rome, 2006), pp. 365–384.

 | DOI 10.1163/9789004330269_008

out any reason belonging to art."[2] In the dedicatory letter written by him in 1571 to the *Illustri e Molto Magnifici Signori Deputati della Fabbrica del Duomo di Milano* (that is, the committee designated to select and to supervise the artists involved in the church's renovation), the architect declares that:

> I will demonstrate (my reasons) not only with my opinion, but also with the authority of the writers and the judgment of architects, and of the most illustrious and famous *perspettivi* of this age; this science, that is the subject of this booklet, is their subject.[3]

As Pascal Dubourg-Glatigny recently noted, Bassi published this treatise in order to solve a practical issue that occurred in the city, invoking for the first time the judgment of his architectural peers whom he defines "experts in this subject."[4]

The political aspects and the final outcome of this quarrel, as well as the consequences on the architect's career, are of no concern for us here; what is interesting to note is Bassi's approach to this particular debate. He uses the names and judgments of different authorities on perspective to support his opinion, referring to ancient authors such as Euclid and Vitruvius while calling for the judgment of modern authorities on perspective such as Witelo, Vignola and Palladio, an attitude that bears a striking similarity to Lomazzo's approach to perspective in his *Trattato*. The artist, in book V on perspective, adopts a similar implementation of the opinions of ancient authors with those of modern *perspettivi* (or, as he calls them, "scholars of perspective"), presenting a discourse on the subject that embeds practical suggestions for

---

2 Bassi, *Dispareri*, p. 29: "Oltre a ciò dico, che niun oggetto di qualsivoglia sorte, o forma non potrà mai con ragione di prospettiva esser digradato, se coll'uno dei due proposti modi del perfetto, o linea piana non si regola. E se così è per qual ragione e regola si è potuto digradare questo smusso segnato DGEF? Lontano, e discosto è certamente dalle buone regole dell'operare, facendosi per se stesso conoscere di non esser fatto con le ragioni dell'arte."

3 Bassi, *Dispareri*, pp. 17–18: "Il tutto dimostrando non solamente col detto mio; ma colle autorità degli Scrittori, e col giudizio degli Architetti e Perspettivi stimati de' più eccellenti, e famosi di questa età; delle quali scienze è materia, e proprio il suggetto è la materia di questo libretto."

4 As Pascal Dubourg-Glatigny recently put it: "Bassi shifts the point of view of the commentary, opening the way to a definition of expertise. In fact, he was one of the first to use the term *esperto* in a treatise, designating one who held practical and proven knowledge, a term which would progressively be distinguished from the term *perito*, which refers to a more theoretical and doctrinal knowledge." (Pascal Dubourg-Glatigny, 2011, <http://architectura.cesr.univ-tours.fr/Traite/Notice/LES0056.asp?param=en>).

painters on "how to make the right perspective" into an exhaustive theoretical discussion based on contemporary and previous optical theories. Moreover, in different parts of his theory of perspective, Lomazzo presents notions that are most likely taken from the writings of people with whom he was familiar through personal connections, as it is the case with the mathematician from Turin Giovanni Battista Benedetti.

At the time the author of the *Trattato* was active as a painter (from around 1550 until early 1570), and later when he started to collect and to organize material for the publishing of his *Trattato*, perspective was a subject widely discussed among artists and architects of the city and beyond, and we have some printed evidence to sustain this claim. Kemp already noted that few books were published solely on perspective in Italy until the mid-sixteenth century.[5] However, besides book I of Alberti's *De Pictura* (translated by Lodovico Domenichi into vernacular in 1547), at least one other example of an author interested in this subject can be found—notably Sebastiano Serlio's *Tutte l'Opere d'Architettura et Prospettiva*. His book III, *Della Geometria, Della Prospettiva,* was published in Venice in 1545. Also, before Vignola's *Le Due Regole della Prospettiva Pratica,* published posthumously by the Bolognese mathematician Egnazio Danti (1583), Daniele Barbaro published *La Pratica della Perspettiva: Opera Molto Profittevole a' Pittori, Scultori, et Architetti* in 1569.[6] A discourse on geometrical constructions of space was therefore thriving in the environments (such as academies) where artists and mathematicians had the possibility to interact, and the need to codify the existent methods of creating the illusion of space was still felt by artists in the second half of the sixteenth century.

Lomazzo dedicates the fifth book of his treatise to perspective, presenting a complex discourse in which optical problems are integrated in the broader discourse on what perspective is, how it should be categorized, and how an artist can appropriate it. He notably divides the subject into three different types: *optica* (which is again divided into physiological and grammatical), *sciografica* (the study of shadows) and *specularia* (concerning mirrors).[7] Apart from the fifth book, the author also takes this subject into account later in the treatise, in the book on practice, ch. XIII (*Delle regole di prospettiva*) and ch. XVII (*Regola*

5 Kemp, *The Science of Art*, p. 69.

6 Iacopo Barozzi da Vignola, *Le Due Regole della Prospettiva Pratica di M. Jacomo Barozzi da Vignola*, edited by Egnazio Danti (Rome: Per Francesco Zannetti, 1583).

7 As Sven Dupré noted, this division comes from the Greek mathematician Geminus. See Sven Dupré, "The Historiography of Perspective and Reflexy-Const in Netherlandish Art," *Netherlands Yearbook for History of Art/Nederlands Kunsthistorisch Jaarboek*, 2011, 61:35–60.

di fare allungare la vista quanto si vuole). In addition to this, we can see that in the later *Idea* three chapters are explicitly dedicated to it: ch. XV, *Delle sette parti della prospettiva*, where he takes into account the ways of making perspectives adopted by his seven governors -Leonardo, Michelangelo, Raphael, Gaudenzio Ferrari, Polidoro da Caravaggio and Titian, ch. XXIII, *Della quinta parte della pittura e della sua spezie*, where he divides the subject into universal and particular perspective, and ch. XXX, *Della via di collocare i corpi secondo prospettiva*. These are all the chapters explicitly dedicated to perspective, but the topic, as we have already noted for the part on color, cannot be legitimately confined to them. Despite the author's constant attempt to divide and categorize rigorously the different parts of painting, this subject and the different issues related to it also occur in other parts of the *Trattato*, especially in the book on light. In this chapter, I attempt to make sense of Lomazzo's discourse on perspective analyzing his division of this subject and the analysis he presents on the power of vision, linking it with the discourse on the eye and of the beams of light. I identify possible sources of information comparing the author's theoretical notions with the practical examples of foreshortening that he quotes in his book. Finally, I present some practical suggestions the painter gives on "how to make a perspective" showing a link with Dürer's discourse and with other Northern figures.

L'Arte della Prospettiva

From the very beginning of the whole Treatise, in the dedicatory letter to Carlo Emanuele, Lomazzo describes perspective as *acutissima* (most insightful), whose effects are "wonderful to see."[8] Already in the prologue he links the discourse on perspective to the discourse on light, which he will develop later in the book, taking into account the illusionistic role of perspective on the connections between the distance of an object and the quantity of light in a painting:

> And if someone should tell me, why do I think that the part (*of a painting*) that is less enlightened, should be nearer to us, since it seems

8 Ciardi, *Scritti*, II, p. 10: "Acutissima è la prospettiva della sua providenza la quale, con si rette linee mirando le tre parti del tempo, produce effetti maravigliosi." (Lomazzo, *Trattato*, dedicatory letter).

> otherwise, that the part with more light in it should be the nearest to us, I will say that it is the art of perspective that does this.[9]

Before book v, the painter also mentions perspective in other places: for example, in the book of proportion, ch. XXIII, when he discusses the proportion adopted for architectural buildings. Here he uses an expression strikingly similar to the one Bassi used in his book, *senza ragione di prospettiva*:

> But to demonstrate this (*how to make the right proportion*) taking as an example the buildings themselves, who does not know that if you place the different *ordini* one on top of the other, according to the proportion in which they should be shown, without any reason belonging to perspective, they would appear too small.[10]

Afterwards, the author mentions perspective several times in the book of Light, as, for instance, in the prologue, where he states that "many painters, skilled in the art of drawing, are considered worthy of praise, but this praise should not be given to them, because they possess no art of perspective."[11] For this reason book v is, as usual, explicitly dedicated to painters. The declared aim of the book is to understand what perspective is and how foreshortening and delineation work, so that artists will not commit mistakes with the eye, which often happens according to the author's opinion, especially when using the *modelli*: "Such as I experienced, trying to make two foreshortenings for figures with the theory of old masters, and the effect was great but nonetheless in practice I found them false, when portraying them using models, with the veil, the *grat-*

9 Ciardi, *Scritti*, II, p. 32: "E se mi dirà alcuno, per qual cagione io giudico che la parte che è manco allumata sia più propinqua a noi, parendo piuttosto il contrario, che la parte più allumata debba stare più verso noi, rispondo che l'arte de la prospettiva fa questo. (*Trattato*, 1584, p. 26).

10 Ciardi, *Scritti*, II, p. 72: "Ma per dimostrar questo in cose più importanti, ne gl'istessi propri edifici, chi non sa che chi facesse gli ordini l'uno sopra l'altro per ordine, secondo la proporzione nella quale s'hanno a mostrare, senza la ragione di prospettiva, farebbe sì che parebbono oltre modo bassi." (Lomazzo, *Trattato*, p. 76).

11 Ciardi, *Scritti*, II, p. 187: "Così vediamo ancora che molti pittori, privi affatto dell'arte del disegno, solo con certa pratica di dare in parte a' suoi lochi i lumi, sono riputati valenti, la qual lode però ragionevolmente non doverebbe esser concessa loro, perchè non hanno né arte de prospettiva, per la quale si vedano nelle fatture loro colorimenti." (Lomazzo, *Trattato*, p. 212).

icola, or the eye. All these ways are not legitimate to make foreshortenings."[12] Despite the earnest statement of the practicality of this book for painters, we soon become aware that, unsurprisingly, Lomazzo does not limit the discourse on perspective solely to what could be practically useful for artists to read in order to learn how to use it correctly. On the contrary, the book presents, along with a reflection on the *vedute* and on the ways of foreshortening, an extended theoretical discussion on the reasons of sight and on the role of the eye in perceiving reality, which mirrors in various aspects some contemporary optical theories circulating in Northern Italy at the time.

The Science of Visible Lines

Lomazzo defines *prospettiva* (ch. III- *Della diffinizione della prospettiva*) as "descending from geometry, the science of visible lines (*scienza delle linee visibili*) and its object of study is the line, of which it seeks the causes, the principles, the universal and immediate elements *per se*."[13] This definition alone indissolubly links the discourse on perspective with the discourse on light and vision, as the "visible lines" are also mentioned in the definition of *lume* (book IV, ch. III). Here, these lines are defined as *linee visuali*, "visual -or visible lines", and Lomazzo states that "these enlightened visual lines are the true subject of perspective, they come to our eye with a pyramidal figure, whose base lays in the thing we see and the angle (or cone of this pyramid) is what comes to our eyes, obtuse and bigger."[14] According to the author, three things make vision possible: the visible lines, the colored body and the power of the eye.

12 Ciardi, *Scritti*, II, pp. 220–221: "Come io ne ho fatto esperienza, approvando due scorti di figure scortate per la via che potevano esser fatte e fondate per l'intelligenza de' maestri, i quali facevano benissimo l'effetto e nondimeno gli ho trovati poi falsi e ritratti da i modelli a pratica, con velo, con graticola, o all'occhio; le quali vie tutte non sono sicure per alcun modo a fare gli scorti." (Lomazzo, *Trattato*, pp. 251–252).

13 Ciardi, *Scritti*, II, p. 222: "La prospettiva subalterna, descendente, & figliuola della Geometria conchiudesi essere scienza delle linee visibili, tal che il suo sogetto è la linea visibile, di cui ella ricerca le cause, i principi, gl'elementi universali primi per se, et immediate; considera il suo genere, le sue spetie et differentie essentiali, e accidentali." (Lomazzo, *Trattato*, p. 254).

14 Ciardi, *Scritti*, II, p. 190: "Le linee visuali illuminate, che sono la propria materia del sogetto della prospettiva, vengono al nostro occhio in figura piramidale, la base della qual piramide sta nella cosa che si ha da vedere, et il cono, o angolo, della piramide e quello che viene al nostro occhio più ottuso e grande." (Lomazzo, *Trattato*, p. 216).

The central role that the author gives to the eye and to the beams of light is quite striking. The optical discourse seems central to his theoretical treatment of the subject and it is later developed in the book, from chapters IV to X, where he discusses the general and particular manners of seeing, the beams of light, the eye, the object and the distances. Apart from the knowledge of Aristotle that the author possessed (which was, as said before, most likely mediated by Italian commentators), he probably also had the chance to discuss these topics directly with his acquaintances. One must have been Giovanni Battista Benedetti, mathematician at the Court of Turin, who published, just one year after the Treatise, the *Diversarum Speculationum Mathematicarum et Physicarum Liber* and whom, as we shall see, the author probably knew personally. The second section of Benedetti's book is entirely dedicated to perspective and it includes a letter entitled *De Visu*, where optical issues related to the power of the eyes are explained in a succinct but effective way.[15]

Aside from the theoretical discourse concerning suggestions on "how to make a perspective", Lomazzo at first discredits the use of models in favor of a more theoretical construction based on the projection and intersection of visual lines. He basically reports and suggests that the reader should use the method of the *quadratura*.[16] However, the author will slightly alter his opinion on the usefulness of models in the later *Idea,* stating that "they could make the bodies known more easily, as well as the shadows and the lights."[17] This change of opinion was possibly due to discussions with fellow artists within the author's circle, such as Aurelio Luini, whom Lomazzo praises regarding his ability in making human proportions, anatomies and landscapes.[18]

The painter declares that the concept of perspective is taken from Lombard sources, since "perspective is what people from Lombardy are most expert of": "Such as ancients and modern *prospettivi* state, especially artists from Lombardy: perspective is to them what drawing is to Romans, color to Venetians,

15 Giovanni Battista Benedetti, *Diversarum Speculationum Mathematicarum et Physicarum Liber* (Turin: apud Heredem Nicolai Bevilacqua, 1585). Both Treatises were dedicated to the prince Carlo Emanuele duke of Savoy. For Benedetti's *De Visu,* see Thomas Frangenberg, "Il *De Visu* di Giovan Battista Benedetti," in *Cultura, Scienze e Tecniche nella Venezia del Cinquecento,* Atti del Convegno Internazionle di studio "Giovanni Battista Benedetti e il suo Tempo" (Istituto Veneto di Scienze, Lettere ed Arti: Venice 1987), pp. 271–282.

16 See Kemp, *The Science of Art,* p. 73.

17 As Giulio Bora already noted, see "La Prospettiva della Figura Umana: gli *Scurti* nella Teoria e nella Pratica Pittorica del Cinqueento" in *La Prospettiva Rinascimentale,* edited by Marisa Dalai Emiliani, Florence 1980, *I*: 295–317.

18 See *Trattato,* book VI, ch. LXII, and book VII, ch. XXIII. Quoted in Bora, Kahn-Rossi and Porzio (eds.), *Rabisch,* p. 362.

and bizarre inventions to Germans."[19] Among the names that he quotes there are the artists Bramante, Vincenzo Foppa and Bernardo Zenale, the last two having written books from which, he claims, Dürer subsequently took up some concepts for his own books.[20] In addition to these names, Lomazzo mentions other artists who wrote about perspective (and especially about foreshortening): Andrea Mantegna and Bramantino, the latter having invented "three ways of perspective", which the author transcribes in the three final chapters at the end of book V.[21]

The two scholars who examined Lomazzo's discourse on perspective in detail disagree on the reliability of his affirmations. While Kemp seriously doubts that the painter borrowed from these "Northern sources" and he declares that these were included probably because of his "chauvinistic attempt" to reevaluate the Northern artistic environment, Bora seems to have no doubt that the artists whom the author quoted must have written on perspective and must have been used as sources by the author.[22] In my opinion the truth lays in between: artists such as Zenale and Foppa were in fact interested in perspective and very possibly wrote about it (writings that were not published in the end), but Lomazzo certainly drew from other contemporary authors, especially when he mentions ancient mathematicians and philosophers, and, as usual, he decides not to quote the names of his contemporary sources. As a proof of this we see that in the book *Della Pratica* the author mentions Foppa's drawings and his *quadrature*, which were a model, along with the (unknown) treatise by Bramante, for later painters such as Raffaello, Polidoro, and Gaudenzio Ferrari.[23] Recent studies have brought Foppa's perspective schemes and drawings to light, showing that the painter, as Lomazzo said, had a strong interest and expertise in the subject.[24] Lomazzo quotes the painter

19 Ciardi, *Scritti*, II, p. 275: "Sì come affermano gli antichi et i moderni prospettivi, massime lombardi de i quali è propria questa parte sì come il disegno è peculiare de' romani, il colorire de veneziani e le bizzarre invenzioni de' germani." (Lomazzo, *Trattato*, p. 317).

20 Ciardi, *Scritti*, II, p. 240: "Nella quale, oltre quello che a di lungo ne scrive, vi sono anco degli schizzi fatti con la penna, si che comprende quasi tutto ciò che ha trattato poi in gran parte Alberto Durero nella sua Simmetria." (Lomazzo, *Trattato*, p. 275).

21 See Lomazzo, *Trattato*, book V, ch. XXII, XIII and XIV.

22 See Kemp, *The Science of Art*, p. 73 and Bora, *La Prospettiva*, pp. 295–301

23 Ciardi, *Scritti*, II, p. 277: "Or quanto alle figure quadrate ne disegnò assai Vincenzo Foppa, il quale forsi dovea aver letto di quelle che in tal modo squadrava Lisippo statuaro antico, con quella simmetria che in latino non ha nome alcuno, e seguendo lui ne disegnò poi Bramante un libro, da cui Raffaello, Polidoro e Gaudenzio ne cavarono grandissimo giovamento." (Lomazzo, *Trattato*, p. 272).

24 See especially Pietro Roccasecca, "Vincenzo Foppa: Evidenze Materiali della Procedura Prospettica nei Tre Crocifissi dell'Accademia Carrara di Bergamo," in *Vincenzo Foppa: Tecniche d'Esecuzione, Indagini e Restauri* (Milan: Skira, 2002), pp. 59–67.

Foppa more than once as a master of perspective, both in the *Trattato* and in the *Idea*. In the latter, the author stresses Foppa's excellence among the Lombard painters in using perspective. Lomazzo quotes practical examples of his production:

> Among them (besides those I have already mentioned, whose names may be found throughout this work) is the noble Milanese Vincenzo Foppa. His works in Milan bear witness to this honor, especially the exceptional illusionistic vault in Santa Maria di Brera, found on the left-hand side, with St. Sebastian tied and surrounded by archers shooting at him. Better than anyone else in this time in Italy, Foppa demonstrated how knowledgeable and informed he was regarding such matters. He naturally deserves first lace for artistic excellence, especially in perspective and in the arrangement of figures, which, in my opinion, constitute all the substace and the foundation of art.[25]

Moreover, in the chapter *Delle distanze* Lomazzo adds that: "Very few understood and speculated on them (distances), and those few did not teach or write about them, with the exception of Vincenzo Foppa, Andrea Mantegna, Leonardo and Bernardo Zenale, whose manuscripts I have seen."[26]

As is well known, Lomazzo certainly had knowledge of at least one of these painters' writings, notably Leonardo's. It seems unlikely that he would have decided to list and mention the other writings in different parts of his books had he not seen them in person, especially given the fact that, in more than one place, the quotations from these authors seem circumstantial, as when he mentions the writings by Bernardo Zenale:

> On this subject, I remember that Zenale, making remarks about various procedures, differed from the opinion of some of his worthy contemporaries in saying that the objects represented far away must be as finished

25 Lomazzo, *Idea*, chapter XXX, translated by Chai, 2013, p. 131.

26 Ciardi, *Scritti*, II, p. 232: "E que' pochi che l'hanno intese e speculate, non le hanno però ad alcuno insegnate nè scritte, salvo Vincenzo Foppa, Andrea Mantegna, Leonardo e Bernardo Zenale, delle cui opere scritte di man loro, oscuramente però, io ne ho assai veduto." (Lomazzo, *Trattato*, p. 264). For other references to Foppa as master in the "arte del far ben vedere," see Ciardi, *Scritti*, II, pp. 199, 200, 353. Other places in which Foppa is declared "excellent in perspective" are in Ciardi, *Scritti*, I, pp. 267, 275, 300, 329.

> and well proportioned as those in the front (...) I read this, along with many other explanations, in certain fragments written in the artist's hand, directed against those who claimed that the smaller the object, the more blurred it should be represented, since nature showed it as such.[27]

This passage testifies to the first-hand knowledge Lomazzo must have had of these unpublished writings that were probably circulating in artists' environments in the city. In addition to this, his statement confirms in passing that even before his *Trattato,* disputes and discussions on the subject of perspective were already taking place among the previous generation of Milanese artists.

The Prologue

In book V, the actual discourse on perspective starts after a long general prologue on what the ultimate goal of painting should be and how an artist should proceed with his judgment. It is interesting to note that this is the only book that begins with a *proemio* before the usual enunciation on the "virtues" of the subject. This prologue clearly enunciates the importance of the role of the eye, the latter being "the scope" and the image being "the medium":

> Now this being the immediate end of this art, it follows eventually that the images are the mean or instrument and the eye is the end, as the first *fondamento* of Aristotle and other philosophers state. And as a consequence the images (*need*) to be proportioned to the eye, which is the immediate end.[28]

These words recall those by Egnazio Danti in his edition of Vignola's *Due Regole* (1583), while referring to his role of commentator for his 1573 edition of Euclid's optics: "Our seeing is possible through the image of the seen thing, which as in

27 Lomazzo, *Idea*, ch. XXX, translated by Chai 2013, p. 130.

28 Ciardi, *Scritti*, II, p. 216: "Ora, essendo cotale il fine immediato di questa arte, ne segue concludemente che le imagini siano mezzo e il fine sia l'occhio, conforme al primo fondamento d'Aristotile e degli altri filosofi posto sopra. E consequentemente, che questo mezzo, cioè le imagini, siano proporzionate all'occhio che è il fine suo immediato." (Lomazzo, *Trattato*, p. 246). As usual, the complex division of this subject makes the author's discourse not always consistent and the categorizations he makes at the beginning are not always faithfully followed throughout the book.

a mirror comes to be in the eye, as Aristotle and the author of this perspective state, and according to the truth, we will demonstrate this with reason and experience."[29]

Lomazzo states here two *fundamenti* or basic rules: first, he takes into account the doctrine of the ancient philosophers who theorized the two means (*mezzi*) of achieving the goal -one which is absolute and the other *ad melius esse* -that is, the best that one can. Second, he states the need for artists to choose wisely between these two means.

He then reaffirms the importance of the senses in making images, since "if you tell me that images do not represent the natural and artificial things, but the intellect and the memory, I will respond that the ultimate goal of images is indeed the intellect, but the immediate goal is the eye."[30]

The role of experience and the link between the eye and the mind are probably taken from Danti's edition of Vignola's *Due Regole*, where the latter quotes Aristotle and his *De Anima.* Lomazzo's theory of vision is further explicated in the later *Idea*, especially in the chapter *Della via di collocare i corpi secondo prospettiva*:

> According to Aristotle, the true vision of lights and objects is performed by an interior sense that apprehends through the eyes the species of colors and objects, both colored and luminous. The vision requires three things: the object, the organ and the medium.[31]

This statement is important in two respects: first, we see that Lomazzo embeds his theory of vision in what he intends to be a practical chapter on how to place bodies according to the rules of perspective. We see that the suggestions in the chapter will shift from Leonardo's *prospettiva aerea* to the method of linear perspective, while at the same time the author mixes the discourse on perspective with that on light and color. In addition to this, one could speculate that Lomazzo's reference to Aristotle in this specific passage reflects his knowledge

29 Vignola and Danti, *Due Regole*, p. 11: "Che il veder nostro si faccia mediante l'immagine della cosa veduta, che come in uno specchio si viene ad improntare nell'occhio, conforme al parere d'Aristotile, e dell'autore di questa prospettiva, e anco alla verità stessa, si dimostrerà apertamente e con la ragione e con l'esperienza."

30 Ciardi, *Scritti*, II, p. 216: "Se mi dici che le imagini non rappresentano le cose naturali et artificiali a l'occhio, ma a l'intelletto et alla memoria, io rispondo e concedo essere il vero che l'ultimo fine delle imagini è l'intelletto, ma l'immediato è l'occhio." (Lomazzo, *Trattato*, p. 246).

31 Lomazzo, *Idea*, translated by Chai 2013, p. 129.

not only of works such as *De Anima*, but also other sources such as *De Coloribus* (as we have seen for the part on color), as the passage bears some similarities with the explanation of what colors are and how they constitute the medium for light presented in the latter:

> So that all colors are a mixture of three things, the light, the medium through which the light is seen, such as water and air, and thirdly, the colors forming the ground, from which the light happens to be reflected.[32]

The prologue then considers the relationship between art and nature, and why the *artifex* must follow the latter as much as he can. Although this concept is quite established in the tradition of art treatises, Lomazzo again puts a striking emphasis on the role of the eye combined with the intellect, which "has to be the rule, the measure, and, in one word, the judge of painting and sculpture." In these words the role of experience sounds similar to Leonardo's theory.[33]

The reason why artists have to learn how to master perspective is explicitly stated in the chapter *De la virtù della prospettiva*; this need for the painter to be "a good mathematician" is repeated in several parts not only in the *Trattato*, but throughout the author's writings, both in prose and in poems. In the *Rime*, for instance, Lomazzo will dedicate a poem to those who are ignorant of perspective: "Those who do not know perspective may as well die, since they cannot draw any thing that from the eyes goes to the mind, and there is no place where it can hide."[34] Certainly the author is here readapting the famous passage by Leonardo:

---

32 Theophrastus (Pseudo-Aristotle), *De Coloribus*, 793b34–794a15. Translated by Hoeppe, *Why The Sky is Blue*, p. 29.

33 Ciardi, *Scritti*, II, p. 217: "Perchè, in somma, l'occhio, insieme con l'intelletto umano, regolato con l'arte della prospettiva, ha da essere la regola, la, misura et, in una parola, il giudice della pittura e della scoltura." (*Trattato*, 1584, p. 246).

34 Lomazzo, *Rime*, II, p. 223:
"*A quelli che non sanno di prospettiva*:
Tanto potria morir quel che non sappi
In prospettiva disegnar niente
Ch'ogni cosa per gli occhi va alla mente
E non vi è luoco fuor donde ella scappi.
Lume et ombra non è ch'ella non grappi
Con le lor gradazioni ch' a la gente
Paion senza parete
Raro è colui ch'in cotal parte incappi."

> Those who fall in love with practice without science are like sailors who enter the river without rudder or compass, and they never know where they are going. Practice must always been built from the good theory, of which perspective is guide and door and without it nothing can be achieved.[35]

This passage was well known and quoted by other writers on perspective, including Danti himself. Perspective is also listed in the "sciences necessary to the painter" (*Idea*, 1590), right after geometry, since "studying perspective, the heart of geometry, the painter will learn about shadows, lights, rays, foreshortenings and ultimately all these aspects that deceive our eyes, making us see what is not."[36]

The subject is here seen as the part of geometry that is necessary for the painter to master, while "geometry teaches the artist about perfect, regular bodies with their proportions and measurements."[37] The importance of geometry and perspective for painters was already clearly enunciated by Alberti in the *De Pictura*, as he declared that "The painter should be as erudite as he can in all the liberal arts: but first of all I want him to be skilled in geometry."[38] From Alberti onwards, different writers on arts stressed the necessity of learning perspective, most notably Leonardo, who, in his *Trattato della Pittura*, declared in the chapter "What the painter ought to know first" that "The apprentice first of all must learn the perspective", as this was the fundamental skill for a painter to create good works of art.[39] The central issues in dealing with Lomazzo's *Trattato* are to determine which kind of perspective the painter is interested in presenting in the book, which rules he actually wishes painters to learn, and what this choice tells us about the general discourse on perspective in Milan at the time. It is reasonable to infer that the author takes into account and discusses a kind of *perspettiva* that distances him from other writ-

35 Leonardo, *Trattato*, p. 54: "Quelli che si innamorano della pratica senza scientia sono come nocchieri che entrano in naviglio senza timone o bussola, che mai hanno certezza dove si vadano. Sempre la pratica deve essere edificata sopra la buona teoria, della quale la prospettiva è guida e porta e senza questa nulla si fa bene."

36 Lomazzo, *Idea*, translated by Chai, 2013, p. 71.

37 Ibidem.

38 Alberti, *De Pictura*, I: "Doctum vero pictorem esse opto, quoad eius fieri possit, omnibus in artibus liberalibus, sed in eo praesertim geometriae peritiam desidero." (<http://www.liberliber.it>)

39 Leonardo, *Trattato*, p. 45: "Il giovane debbe prima imparare prospettiva."

ers on art. Lomazzo, for instance, does not seem to follow the discourse on the *perspettiva aerea* presented by Leonardo, or the discussion on the *punto focale* presented by Alberti, although he is aware of their writings as he mentions both authors more than once describing them as the authorities from whom everybody learns. Throughout book V, the painter instead uses a combination of different sources to offer a mathematical construction of perspective, which is placed among the chapters dedicated to a detailed discussion of the role of the eye and on the rays of vision. It links the theoretical discourse on what perspective is to previous and contemporary theories of vision. In addition to this, Lomazzo assembles an impressive amount of material on the different points of view or *vedute*, revealing a renewed interest in architecture and foreshortening. This interest can be explained by contextualizing the material that he presents with the pictorial commissions he received (especially for the vault of the Cappella Foppa in San Marco), and acknowledging the centrality of the discussion on architectural *decorum* in the city of Milan at the time. This is illustrated by quarrels such as the one mentioned earlier between the architects Bassi and Tibaldi. In fact, this quarrel was well known by Lomazzo. In order to judge it the *Prefetti* of the Duomo had called upon one of his acquainances, the architect Giacomo Soldati. Soldati was also acquainted with the mathematician Benedetti and was also a member of the Blenio Academy under the name *Compà Soldarogn*.[40]

The Division of *Prospettiva*

After defining *prospettiva subalterna* as the science of visible lines, which searches for the causes, the principles and the universal elements, Lomazzo makes a complex division of the subject which he struggles to follow throughout the book.

The division can be summarized in the following diagram (fig. 5.1). The first part (the "optical perspective") is the one that the author develops in detail, and the only one that he actually discusses in book V. A short paragraph is dedicated to the second kind, the *prospettiva sciografica* or the science of shadows, in which the author states that its rules can go under the *grammica* part, as "it considers the shadows, causes, principles, elements, differences, *specie*,

40 See Morigia, *La Nobiltà di Milano*, p. 476 and Isella, *Rabisch*, p. 367.

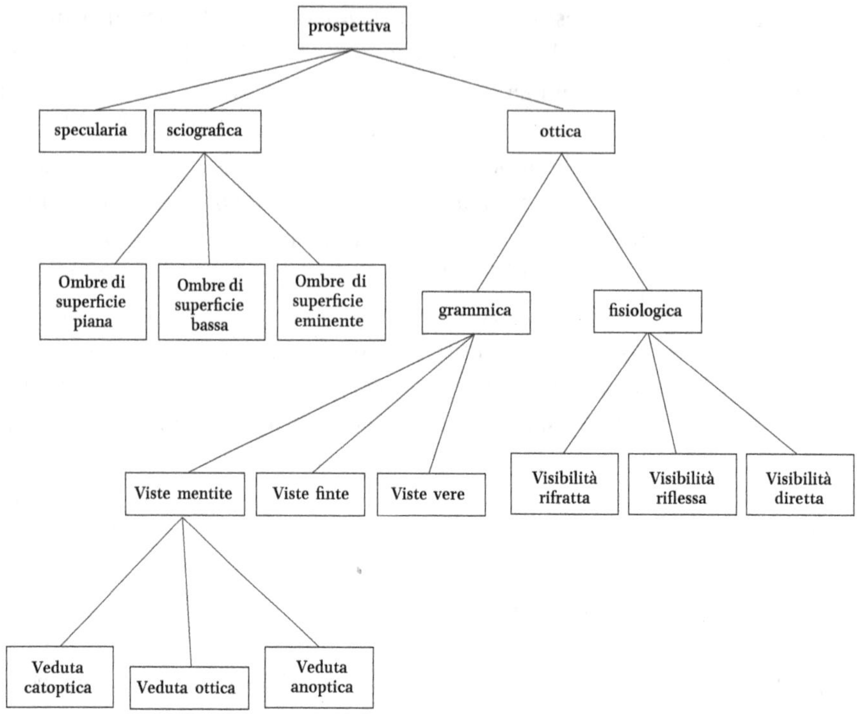

FIGURE 5.1 *Lomazzo's division of* prospettiva

parts and existential passions (…) and all its reasons are under the *grammica*."[41] Lomazzo then admits that some people can compare it and think of it as the *scenografica*, which Vitruvius talked about (and whom the author probably knew through Barbaro's *Pratica*), but he then cuts the story short by saying: "Anyone can think of it as it pleases him, I will follow the ancient definition and division of perspective."[42]

41 Ciardi, *Scritti*, II, p. 224: "La seconda spezie, detta sciografica, tratta compiutamente delle ombre, cause, principii, elementi, differenze, spezie, parti e passioni essenziali; e rende le cause delle varietà vedute delle imagini delle cose, col mezzo delle distanze, lontananze, vicinità di siti, sopra, sotto e mezzo." (Lomazzo, *Trattato*, p. 254).

42 Ibidem: "Ma intendala ciascuno come vuole, io seguirò il detto ordine e la vera et antica difinizione e divisione della prospettiva." On this see Riccardo Bellé, "Alcuni Aspetti dell'Ottica di Giovanbattista della Porta: dalla 'Magia Naturalis' al 'De Refractione'," paper at the conference *Lumen, Imago Pictura*, at the Biblioteca Hertziana, Rome, April 2010.

Regarding the third part, that is the science of mirrors or *specularia*, the painter takes into account the reflection of rays of vision with the help of mirrors. He infers that this part of perspective "shows all the tricks (*inganni*) of the mirrors, which are different according to the mirror's shape: concave, round, plain, pillar-like, pyramidal, swelling, angular, inverted, erected, regular, irregular, solid and clear", a division that can already be found in Witelo and was taken up by another source for Lomazzo, namely Giovan Battista della Porta.[43]

In the *prospettiva ottica*, which is divided into *fisiologica* and *grammica* (that is, the art of delineation), Lomazzo writes two chapters on what sight is and on the rays of vision that depart from the eye, continuing with a detailed account on the three different types of sight (from below, on the line of the horizon, and from above). On these three types, which highlight the author's interest in the right foreshortening, he declares that: "Many painters, architects and sculptors were skillful in making these lines", and among others he quotes the aforementioned Martino Bassi.[44] The division of the *viste mentite* is further elaborated upon later in the book, and is not confined to a discussion on these three *modi di vedere*. Lomazzo instead divides the *mentite* into six parts, using one or more works of art to elucidate each part.[45] The author's more practical discussion on the division of the sights, which are *anoptica, optica* and *catottica* can be found in ch. x, xi, and xii.

As Panofsky already pointed out, this division of perspective into three ways of sight can be found in the contemporary Codex Huygens attributed to Carlo Urbino. Urbino also divides the three *vedute* into: the normal view, the worm's view, and the bird's view (fig. 5.2).[46]

The codex, written around 1570 (when Lomazzo probably started to gather material for the *Trattato*), dedicates the fifth book, as our author does, to the subject of perspective. Notably, many of the ideas presented there derive from Leonardo's lost sheets, reporting different methods of foreshortening and eventually suggest the adoption of the distance which is "proportionate and it

43 Ciardi, *Scritti*, ii, p. 224: "Reflessione de i raggi, e porge aiuto al artificio de gli specchi, mostrando tutte le affezioni e gl'inganni di quelli, che diversamente si veggono, secondo le varie forme loro: incavate, rilevate, piane, colonnari, piramidali, orbinati, gobbi, rotondi, angolari, inversi, eversi, regolari, irregolari, sodi e chiari." (Lomazzo, *Trattato*, p. 254).

44 Ciardi, *Scritti*, ii, 225: "Nelle quali sono ora molto pronti, tra gli altri pittori, scultori et architetti, come il Clariccio il Meda col Bassi." (Lomazzo, *Trattato*, p. 255).

45 See Lomazzo, *Trattato*, book v, ch. xiii–xviii.

46 See Panofsky, *Codex Huygens*, p. 100 ff, quoted in Ciardi, ii, p. 233, n1.

is comfortable to the eye"[47] (fig. 5.3). In addition to this, we can easily see how these last three chapters in the book of perspective (ch. x, xi, xii) recall the final four chapters in the book of light, that is the parts dedicated to the *sciografica* (book iv, ch. xxii, xxiii, xiv, xv).

It is in fact in the book of light and not in the book on perspective that Lomazzo discusses the theory of shadows according to the three ways of sight, stating that shadows "when they are over our eyes according to the *anottica* view they need to make the figure foreshortened and the lights and shadows must follow the lines", while in the *ottica* view not much shadow is needed.[48] In the last view, that is the bird's view, the author suggests the proper dark color to use, linking the discourse on light to that on colors (book iii) and again to the general issue of *decorum*.

The Reason of Sight and Lomazzo's Connections

In the fourth chapter on the reasons for sight in general, Lomazzo makes a list of ancient philosophers who discussed what makes vision possible and how the power of the mind relates to the eye, recording their names and summarizing the opinions of each:

> Among the best philosophers, which I recall to have read concerning the reason of our sight, I find diverse opinions. For Plato thinks it is caused from that brightness which proceeds from the eye, whose light passing through air meets with the one, which is reflected from the bodies. Now that light with which the air is enlightened from the sun disperses itself and turns into the virtue of sight. And this is also Galen's opinion. (…) But Hipparchus believes that the beams, which issue from the eye reaching into the bodies, deliver to the sight the things they have received. Aristotle thinks that the incorporeal similitude and qualities of things come to the sight through the alteration of the air, which surrounds the visible things. Porphyry teaches that neither the beams nor the resemblances nor any other thing are the cause of sight, but only the mind itself that knows all

47 Panofsky, *Codex Huygens*, f. 96 r.

48 Ciardi, *Scritti*, ii, p. 213: "Ma per venir alle ombre sopra il nostro occhio, cioè nella vista anottica, dico che quanto più si veggono le figure scortare e le parti interiori inalzarsi e quella abbassarsi che i lumi e le ombre andando dietro alle linee." (Lomazzo, *Trattato*, p. 243).

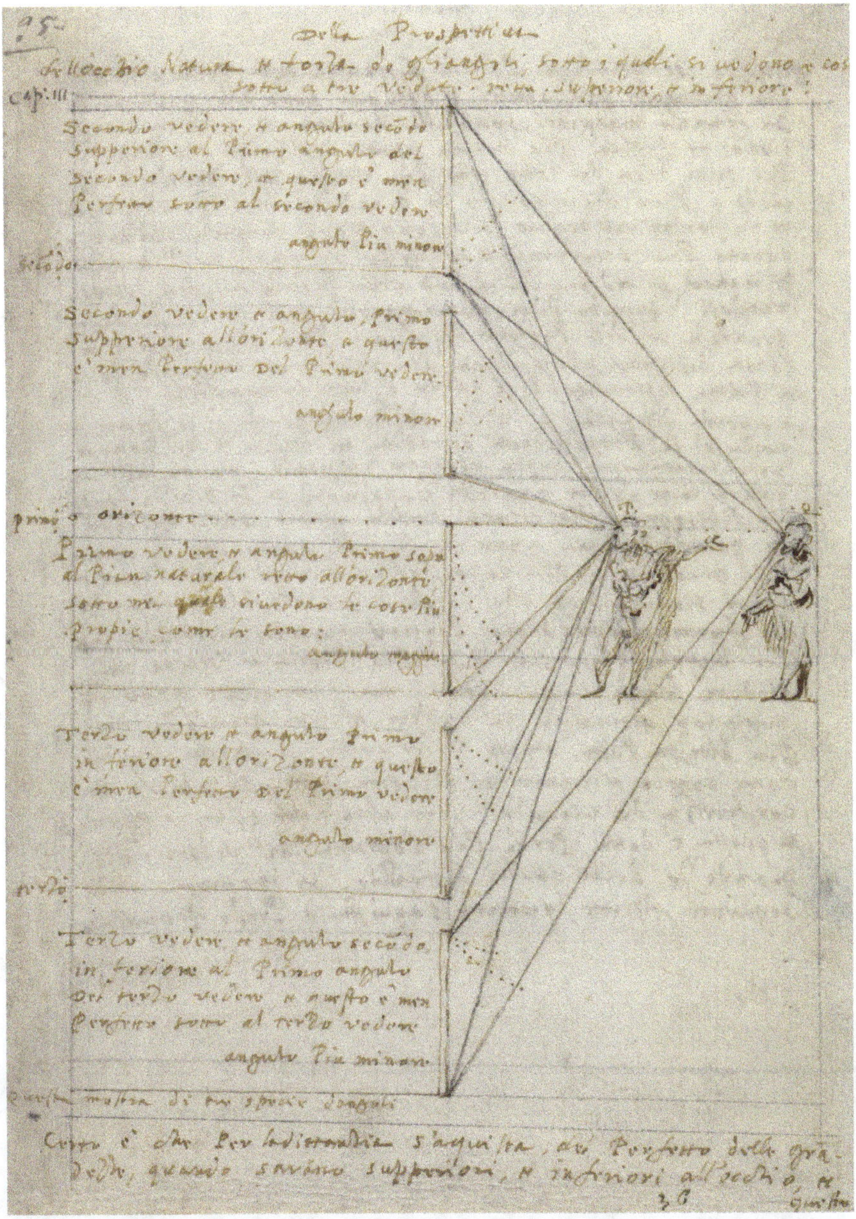

FIGURE 5.2 *Carlo Urbino,* On the three ways of seeing, *Codex Huygens, fol. 95, 1580 ca, New York, Pierpont Morgan Library*

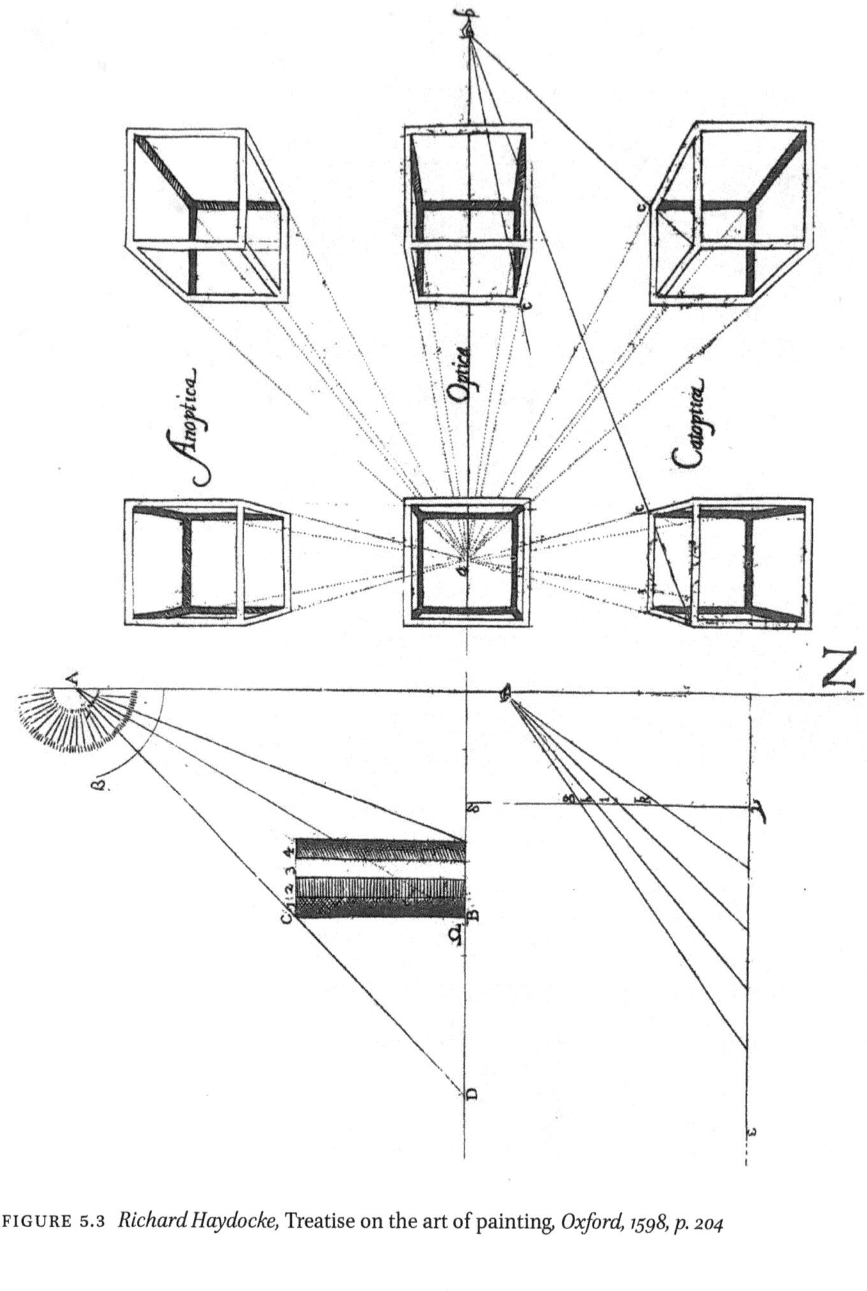

FIGURE 5.3 *Richard Haydocke,* Treatise on the art of painting, *Oxford, 1598, p. 204*

> the visible things and is known in every thing. The Geometricians and *perspettivi* draw near Hipparchus' opinion, they imagine certain cones meeting with the beams, which proceed out of the eye, whence the sight comprehend many visible things together (…) Alchindus is of another opinion, while saint Augustine states that the power of the mind works in the eye.[49]

According to Ciardi, the author takes these names from Danti's edition of Euclid, and in particular from this passage:

> We must follow Euclid's opinion, and Plato's, his teacher, which is followed by all ancient mathematicians, such as Alchindus, Heliodorus of Larissa and Theon of Alexandria, and Galen, in the seventh book of Ippocrat's precepts, and of Plato's, in the second part of his treatise on the eyes.[50]

Most probably the author knew this passage well. However, one can easily see when comparing the two texts, that the discourse on vision and, in general, the discourse on perspective that Lomazzo presents are far more developed and

49 Ciardi, *Scritti*, II, pp. 225–226: "Per quello ch'io mi ricordo d'aver letto, circa alle ragioni del vedere, appresso degl'eccellenti speculari, diverse e varie sono in ciò l'opinioni et i pareri. Perciò che Platone crede che la vista si faccia secondo la chiarezza cioè quella che viene degl'occhi, scorrendo la luce ad uno aere estrinseco, e quella che è rivoltata da i corpi incontrando la luce. Ma quella che sta circa l'aere di mezzo ha faccia che si sparge e si rivolge alla virtù del vedere. Del qual parere è anco stato Galeno; (…) Ma Ipparco dice che i raggi distesi da gl'occhi, toccando quasi con una certa palpitazione sino a quelli corpi, rendono quel che pigliano alla vista. Gli epicurei affermano che le sembianze delle cose che appaiono da se stesse entrano ne gl'occhi. Aristotile è d'opinione che le simiglianze, non già corporee, ma secondo la qualità, per la alterazione dell'aere, il quale è nel circuito delle cose visibili, viene sino alla vista. Ma, Porfirio dice che ne i raggi, ne le sembianze, ne alcuna altra cosa e cagione del vedere, ma è l'istessa anima che conosce se medesima visibile e si conosce in tutte le cose che sono. I geometri e prospettivi, accostandosi a un certo modo ad Ipparco, sottoscrivono certi coni fatti all'incontro de i raggi, i quali si mandano fuora per gl'occhi, onde la vista comprende insieme molte cose visibili, ma, certissimamente quelle dove i raggi s'incontrano insieme. Altro dice Alchindo degl'aspetti; santo Agostino tiene che la potenza delle anime faccia alcuna cosa nell'occhio." (Lomazzo, *Trattato*, p. 256).

50 Ciardi, *Scritti*, II, 226, n3: "Dobbiamo seguire la sua (*Euclide*) opinione, e di Platone suo maestro, a la quale aderiscono tutti i matematici antichi, come Alchindo, Eliodoro Larisso e Teone Alessandrino, e Galeno, nel settimo libro dei precetti di Ippocrate, e di poi Platone, e nella seconda parte del trattato degli occhi."

complex, while the sources from which he borrows cannot be legitimately confined to the Euclidian tradition mediated by Danti. It is probable that the author had different writings in mind by other mathematicians and *perspettivi* with who he was personally in contact.

As to whom he is referring, we have a useful hint in the later *Idea*, where Lomazzo gives a list of treatises by modern mathematicians that a painter should know in order to understand "proportions and all the other things":

> They (*painters*) will not lack clear and beautiful treatises by modern mathematicians from which they can learn and they can take the true proportions and everything else, such as the treatises by Torriano, Petrus Apianus of Ingolstadt, Nostradamus, Cardano, Moleto, Ottonai, Tartaglia, Commandino, Benedetti, Pigliasco, Siglio, Giuntino and Baldini.[51]

Although it is common for the author to make lists of names without further elaboration on them, we know that he personally knew or at least was aware of the writings of three among the names mentioned: Girolamo Cardano, Bernardino Baldini (mathematician and member of Blenio under the name of *Compà Baldign*), and Giovanni Battista Benedetti (1530–1590), astrologer and mathematician at the court of Turin. As we have previously discussed, it is not certain whether Lomazzo actually visited the court of Turin during his travels around Italy in his early years. However, it is known that Benedetti made the horoscope of the author, as the latter states in the poem dedicated to him:

> From philosophy comes the prudence and the knowledge of those intellects. (…) Of these the wise and rare to the world Benedetti takes a great part. He cannot be compared to anyone, such its valor is so sublime. And I rejoice that he liked my paintings so much that he wanted to set the time and place in which I was born.[52]

---

51 Lomazzo, *Idea*, translated by Chai 2013, p. 100.

52 Lomazzo, *Rime*, III, p. 158: "Da la filosofia nasce e discende/La prudenza e il saper degli intelletti/Coi quali essendo nel dispor perfetti/A ognuno suo diritto e sua ragion si rende/Di questa sì gran parte se ne prende/Il saggio e raro al mondo Benedetti/Che d'agguagliarlo in vano è chi s'affretti/Tanto sublime suo valor s'estende./Però tanto godo io che si gli piacque/La mia pittura e che perciò egli volse /L'ora et il punto nel qual nacqui al mondo."

Benedetti was a keen scholar of perspective, and as we have said, he dedicated the second section of his *Diversarum Speculationum* to a discussion on the subject (*De rationibus operationum perspectivae*). In this part of the book, which was published in 1585 and dedicated to Carlo Emanuele of Savoy, he declares his intention to understand "the inner and true causes of perspective."[53] He claims that this book has been inspired by discussions and talks with contemporary artists. Moreover, the seventh chapter of this section on perspective is dedicated to "Iacobo Soldato Mediolanensi Serenissimi Ducis Sabaudiae Architecto Peritissimo", that is the architect and engineer Giacomo Soldati mentioned earlier in the chapter. It was he who was called to judge the quarrel between Bassi and Tibaldi, and additionally, was a close friend of Lomazzo and a fellow member of the Blenio academy.[54] Soldati was an engineer and the military adviser to the Duke of Savoy, working both in Milan and Turin, and Benedetti addresses him at the beginning of the chapter, stating that: "Not so long ago you and I had many discussions on perspective, so that our mind is totally focused on it."[55] This passage sheds light on the centrality of the topic during the discussions between the mathematician from Turin, the Milanese engineer, and the unknown artists whom Benedetti generally refers to (perhaps the artists belonging to the academy of Blenio) in his text.[56]

This network of people gives us a glimpse of the reasons why we find topics such as architectural, mathematical and optical problems in Lomazzo's treatise and especially in book V.

---

53 Benedetti, *Diversarum Speculationum*, II, p. 119: "Veras internasque causas operationis perspectivae." Benedetti states there, as Lomazzo does, that he wants to talk about this subject because who does not know perspective will commit many errors: "Qui huiusmodi operationis regulas prescribunt, cum ipsius effectuum veras causas ignorent, varios diversosque errores committurit." On Benedetti as astrologer at the court of Turin see Pietro Daniel Omodeo, "La Cosmologia Infinitistica di Giovanni Battista Benedetti," *Bruniana & Campanelliana*, 2009, *XV*, *1*:181–190. For Benedetti's perspective, see J.V. Field, "Giovanni Battista Benedetti on the Mathematics of Linear Perspective," *JWCI*, 1985, *XLVIII*:71–99.

54 Benedetti, *Diversarum Speculationum*, ch. VII, p. 133. See Field, "Giovanni Battista Benedetti", p. 92: "Unfortunately, Wittkower's brisk comment that not much is known about this Giacomo Soldati still seems to be substantially justified."

55 Benedetti, *Diversarum Speculationum*, II, p. 133: "Superioribvs diebus non diu postquam de perspectiuis inter nos sermonem habuimus, dum animus totus adhuc in his esset."

56 For the dates of Soldati's work in the two cities, see Field, "Giovanni Battista Benedetti", p. 93.

The author's voice and explicit writing strategy for the reason for sight can be found at the very end of the fourth chapter, where he states the way in which he wants to discuss vision: "I will join these opinions (the opinions of ancient authors) with the opinions of others in the following chapters, according to those which seem closer to truth, and I will handle these opinions freely and as painters do (*da pittori*), least some choleric fellow (…) might think that I speak randomly."[57]

What he declares in this part of the book is his intention to discuss the theory of vision in a manner understandable for painters. His use of long lists and his division of the book into chapters seems to reflect the urge to create a "didactic" book.

In the chapter "On the reasons of sight in particular", the discourse switches to the eye thus defined:

> The instrument of sight, which has many coats or skins in the middle whereof, is the sight, which rises from a certain straight passage called *ottero* til the extremity of the pupil, and it comes from the brain. From this comes the visual virtue (*virtù visiva*), and as the beams come forth they dilate, because they come out with great power and thickness.[58]

It is interesting to compare this definition with the definition of the organ that the author gives later in the XXX chapter of the *Idea*:

> The visual organ is the eye, attached to the bifurcated nerve of optical colors that stretches from the brain to the pupil of the eye, where the visual ability of the nerve permits the reception of the image or form

57 Ciardi, *Scritti*, II, p. 226: "Io, accompagnando questo parere con gl'altri, ne' seguenti capitoli particolarmente, secondo che più pareranno vicini e conformi alla verità, ne tratterò alla libera e da pittori, acciò che alcuno stitico (…) non pensasse ch'io parlassi fuori di figura probabile." (Lomazzo, *Trattato*, pp. 256–257).

58 Ciardi, *Scritti*, II, pp. 226–227: "Primieramente l'occhio, istromento del vedere, ha più spoglie, et in mezzo è il vedere, il quale riesce per uno contratto, chiamato ottero, insino all'estremo della pupilla, e viene dal cervello. E per quello viene la virtù visiva, e come arriva fuori, i raggi si dilatano, perchè escono fuori con grandissima possanza e spessezza." (Lomazzo, *Trattato*, p. 257).

> from the crystalline liquid in the pupil. This image is brought to the *sensus communis*, where the difference between colors is judged.[59]

Whilst the first definition presents a more "functional" understanding of the organ of sight that is linked to the beams of vision (following the Euclidian concepts commentated by Danti), the second definition, which we could define as more "anatomical", links the anatomy of the eye to the power of discerning colors thanks to the *umido cristallino*, which is probably taken from Vignola's *Due Regole*, more specifically from his *definitione quarta* entitled *Centro dell'occhio è il centro dell'humore cristallino* ("The centre of the eye as the center of the crystalline").[60]

Lomazzo's make two distinct definitions while Danti, in his work, eventually sums up both the functional and the anatomical ones in a single comprehensive definition:

> Therefore Aristotle says that the species or image of the thing seen spreads out in the air so that it arrives within our eye to impress itself in the *humor crystallinus* in which vision primarily occurs even though the entire substance of the eye has its part in vision.[61]

After the author's definition of what the eye is, he starts a more thorough discourse related to the beams that come from it, which also seems to recall the Euclidian tradition mediated by Danti. As Frangenberg already pointed out, Danti's remarks, which constitute "a typical exposition of mainstream Cinquecento thinking on optics", aimed at explaining Vignola's treatise while "drawing upon different and diversified traditions -geometrical, medical, physiological and philosophical."[62] Lomazzo seems to borrow the optical concepts that are of more interest for him from Danti, as one can see in ch. VI, *De i raggi del vedere.*[63] Here the artist quotes Euclid twice: first to relate the different angles

---

59 Lomazzo, *Idea*, translated by Chai, 2013, p. 129. "L'organo del vedere è l'occhio, al quale si distende il nervo de' colori visivi biforcato dal celabro insino alla pupilla dell'occhio, ove la virtù visiva nel nervo contiene l'idolo, avera forma dell'umido cristallino che è nella pupilla dell'occhio et è portato al senso commune, ove si fa giudicio della differenza de' colori." (Ciardi, *Scritti*, I, p. 326).

60 Vignola-Danti, p. 2.

61 Vignola-Danti, p. 12, in Thomas Frangenberg "Egnatio Danti's Optics. Cinquecento Aristotelianism and the Medieval Tradition", *Nuncius: Annali dell'Istituto e del Museo di Storia della Scienza di Firenze*, *III*.1:3–38, p. 16, n52.

62 Frangenberg, "Egnatio Danti's Optics.", pp. 7–8.

63 Ciardi, *Scritti*, II, pp. 229 ff. (Lomazzo, *Trattato*, pp. 259–261).

of sight according to which we perceive the different sizes of objects, and then to quote his *terza supposizione* in order to explain what foreshortening is and how perspective lines should pass through an object in order to construct it: "But what cannot be seen is, as Euclid states in the third supposition, what one can barely see, I am talking about the visual objects, which we foreshorten with formal lines."[64]

The discourse continues with a reflection on the ability of the soul to perceive reality of Aristotelian flavor, which probably inspired the prologue to Lodovico Cardi's *Trattato Pratico di Prospettiva*: "Our soul is contained in the body and it can feel nothing of the external things by itself, only through the senses."[65] These words recall the Aristotelian discourse on the soul and its connection with the senses, a discourse similar to what we find in Lomazzo's prologue and later in this chapter: "Our soul in itself does not produce these effects, but one can only let it have the things that are ordered by it, such as seeing, walking and similar."[66] Though it is difficult to pin down the direct sources for this and other passages of Aristotelian derivation, it is clear that the author takes up the connection between the mind and the eye which makes vision possible. Like other writers such as Alberti or Barbaro, he is not much interested in defining whether the beams come from the eye to the object of sight or vice versa. In the *De i raggi del vedere,* he takes for granted that beams depart from the eye (as he states: "The rays of seeing are those which depart from the eye and catch all the details of the objects that are to be painted.").[67] Instead, as we have already seen in the prologue, he is much more interested in determining how the object appears to the eye according to the different distances (ch. VIII, *De le distanze*), since "such experiments of painters concerning this matter, who have considered that without distance they can make all objects with reason, and then who could find out this distance, are important in order

64 Ciardi, *Scritti*, II, p. 229: "Ma quello che non si può vedere è, come dice Euclide nella terza supposizione, quello che appena si può vedere, parlo delle cose visive, che con linee formalmente s'introducono a doversi scortare." (Lomazzo, *Trattato*, p. 260).

65 Lodovico Cardi, *Trattato Pratico di Prospettiva di Ludovico Cardi Detto il Cigoli: Manoscritto Ms 2660 A del Gabinetto dei Disegni e delle Stampe degli Uffizi* (Rome: Bonsignori, 1992), p. 25.

66 Ciardi, *Scritti*, II, p. 227: "L'anima nostra in sé non ha questi vari effetti, ma è sufficiente a fargli aver ad altri le cose ordinate in lei, come vedere, andare e simili." (Lomazzo, *Trattato*, p. 258).

67 Ciardi, *Scritti*, II, p. 228: "I raggi del vedere che sono quelli che, partendosi da l'occhio vanno pigliando tutte le particolarità de gl'oggetti che si vogliono dipingere." (Lomazzo, *Trattato*, pp. 259).

to judge which paintings are beautiful." He includes in these words a practical aesthetic judgment.[68]

What we have here is the opinion of a painter, who after discussing the theories of vision with contemporary men of letters, tries (at least on paper) to apply them to the realm of his workshop and who, at the same time, wants to build his own theory of perspective from the realm of his painterly practice. In the next section, we shall see how Lomazzo applies the theoretical discourse on foreshortening and fictitious views giving practical examples of works of art that he had personal knowledge of.

The *Viste Mentite*: a Comparison with Paintings

After the theoretical chapters on vision, Lomazzo's discussion on the *viste mentite* comes forth in the book of perspective quite unexpectedly, when he begins a section by creating a new categorization that he did not mention in the introduction of the book. This might be an indication that this part was actually added at a later stage. The *viste mentite* are thus subdivided: *prima suprema perpendicolare, seconda obliqua, terza superiore, quarta mezzana, quinta inferiore, sesta intrante.*[69]

As different scholars already noted, it is somewhat challenging to understand what Lomazzo means by "fictitious views" in these chapters given the fact that he does not provide illustrations to exemplify his definitions.[70] However, one may find it helpful to compare these *viste mentite* with the works

68 Ciardi, *Scritti*, II, p. 232: "Tali esperienze fra i pittori che sopra di ciò hanno considerato che senza quella si possano formare tutte le cose e tuttavia paiono giuste e fatte con ragione, e trovare questa distanza detta di sopra, come più rara e più bella in tutte le opere, e conoscere per questo dove la si trova e perciò giudicare quali siano le opere belle." (Lomazzo, *Trattato*, p. 264).

69 Lomazzo, *Trattato*, ch. XIII–XVIII. For a recent discussion on these *viste mentite*, see Dubourg-Glatigny, "La Question de la Perspective Matérielle", especially pp. 376–384.

70 On the subject of Lomazzo's *viste* compared with the ones presented in the Codex Huygens, apart from the studies of Bora already quoted in this chapter, see also Stefano Marconi, "La Proiezione della Figura Umana nelle *Regole del Disegno* del Codex Huygens," in *Prospettiva*, 1998, pp. 183–189. Marconi's insightful analysis of the Codex allows to interpret the aims of the Codex's author in synthetizing schematically the different angular *perspettive*.

of art, mainly taken from the North of Italy, that he lists as examples for each of these views, focusing naturally on the ones which we still have today.[71]

The first *vista mentita suprema perpendicolare* includes those paintings that are on flat surface but viewed from below, "in such a way that this figure seems larger than higher, but the foreshortening make us see it such as it would be in reality."[72] This type is usually found in the depiction of saints in the corners, and as an example he mentions the fresco portraying the Holy Father from the *tiburio* of Santa Maria di Campagna in Piacenza by Pordenone (fig. 5.4).[73]

Lomazzo certainly knew this fresco personally, since he worked in the city around 1567 to make a fresco for the monastery of Sant'Agostino, the already mentioned *Cena Quadragesimale* now lost.[74]

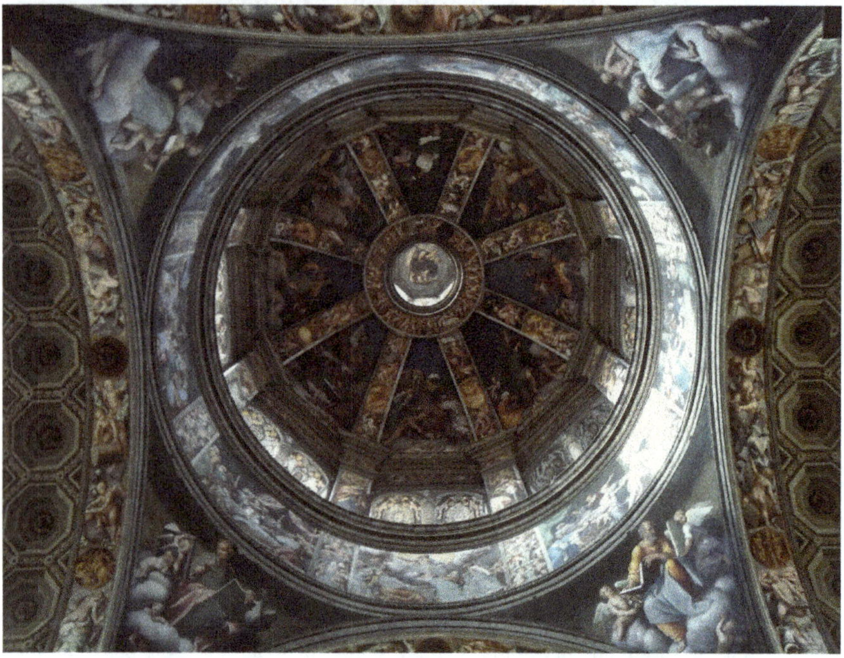

FIGURE 5.4 *Pordenone,* The holy father, angels, prophets and sybils, *c. 1550, Piacenza, Santa Maria di Campagna*

71 See Panofsky, *Codex*, p. 101.

72 Ciardi, *Scritti*, II, p. 235: "Ma quelle che sono al longo, per lo più si scortano di maniera che questa tal figura si dimostra più larga che alta, et opera dentro questa meraviglia, che la ci fa parere grande come se così veramente fosse." (Lomazzo, *Trattato*, p. 270).

73 Ibidem.

74 Only a preparatory drawing survives in the Windsor Castle (fig. 5.13). The fresco was destroyed during World War II, in 1943.

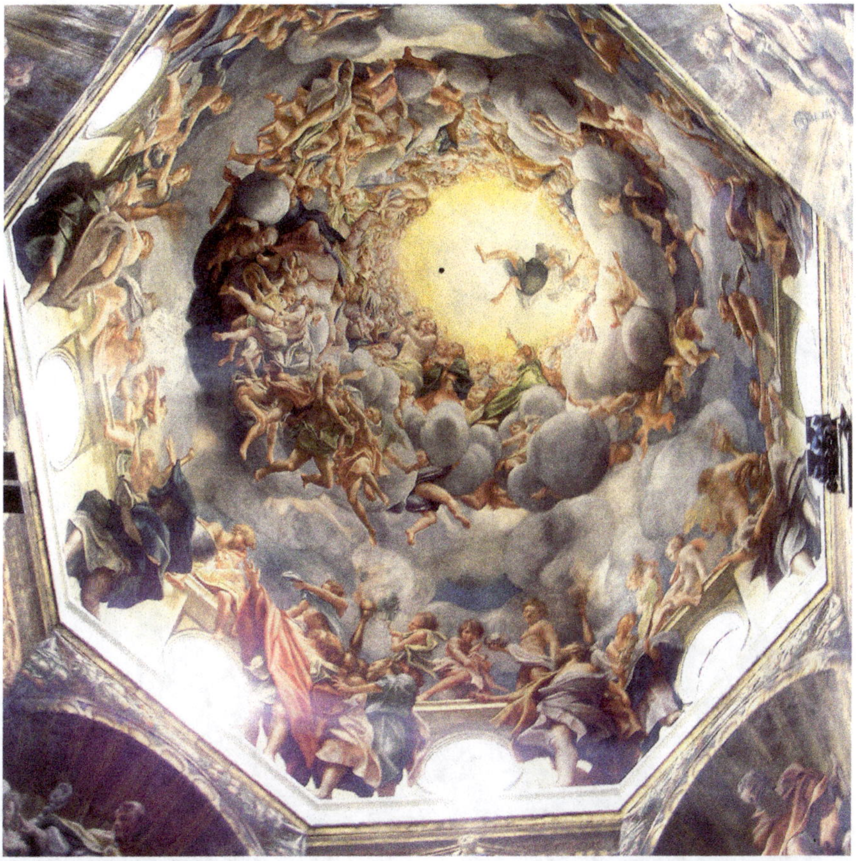

FIGURE 5.5 *Correggio,* Ascension of the virgin, *c. 1530, Parma, Santa Maria Assunta*

The second *vista* is oblique, "which is the most difficult to achieve, since one needs to be careful not to make mistakes."[75] It is the type of foreshortening that can be found in chapels, vaults and in other circular surfaces. As an example, the author takes the *Ascension of the Virgin* in Parma by Correggio (fig. 5.5).

It is likely that Lomazzo was inspired by this fresco when painting his *Glory of Angels*, one of his few surviving frescoes in the Cappella Foppa in San Marco, Milan (fig. 5.6).

From this fresco we also have one famous preparatory drawing, which shows the *vista mentita* from below, at the University Museum in Princeton (fig. 5.7). From this drawing, one can also infer that Lomazzo might have actually used

75 Ciardi, *Scritti*, II, p. 235: "Questa via di scortare è la più difficile." (Lomazzo, *Trattato*, p. 270).

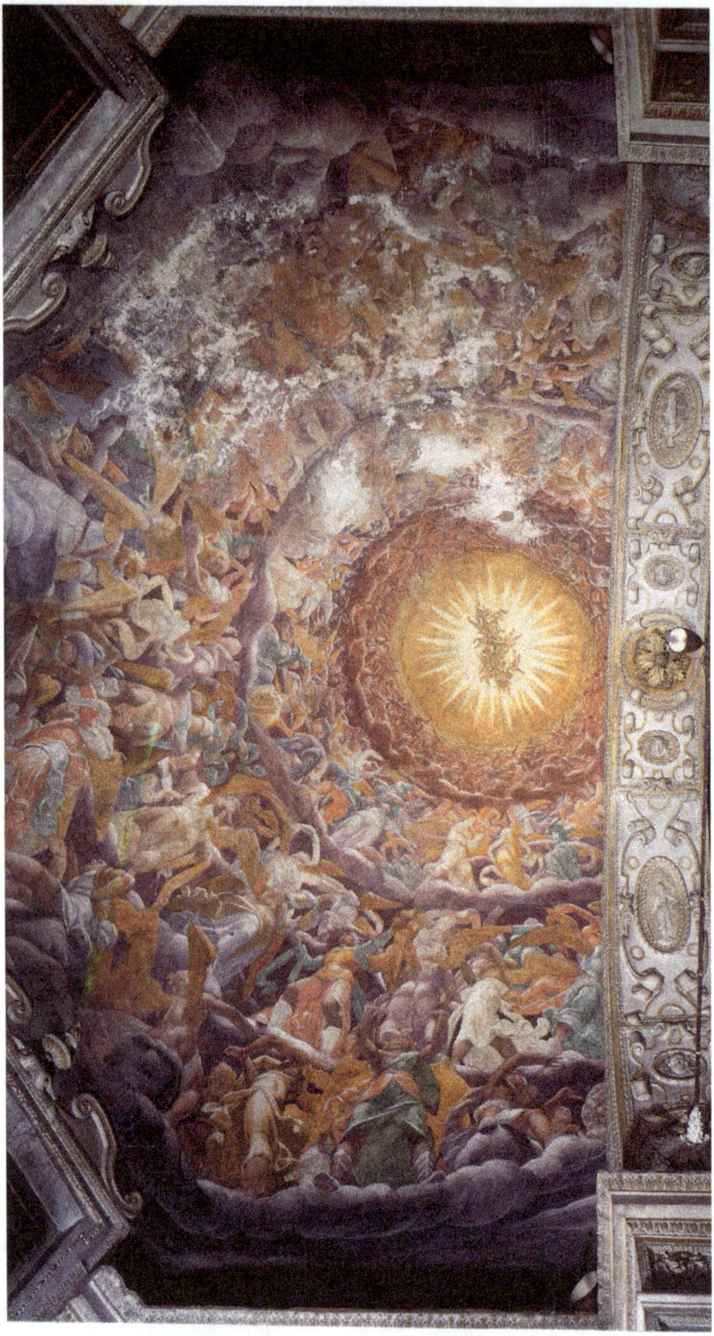

FIGURE 5.6 *Giovanni Paolo Lomazzo,* Angels in glory, *c. 1570, Milan, San Marco, Cappella Foppa*

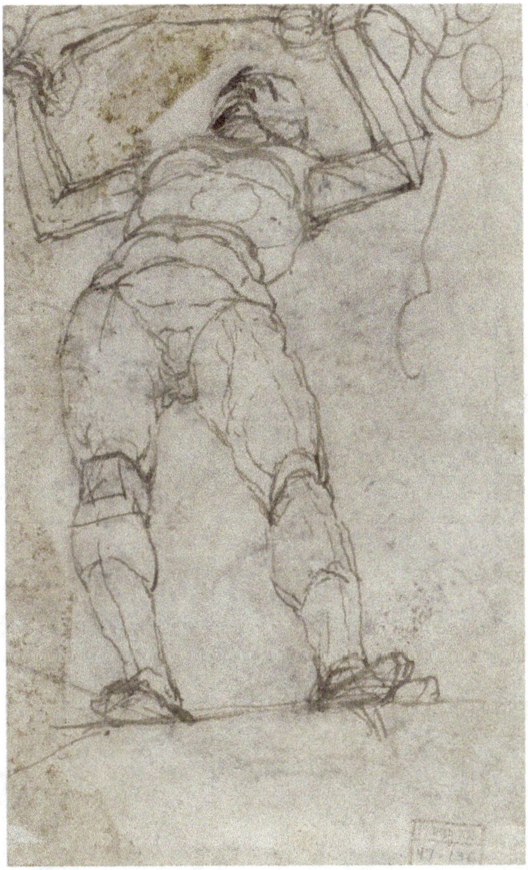

FIGURE 5.7
Giovanni Paolo Lomazzo, Study of a figure, *c. 1570, Princeton, University Art Museum*

models in his workshop to make the right perspective, as the drawing seems to illustrate what could be a clay model, which can be especially noted in the limbs of this figure.[76] The same view can be seen in a folio of the Codex Huygens (fig. 5.8).

The subsequent three *viste* (third, fourth, fifth) are analyzed taking into account only vertical surfaces and following the three *modi di vedere* (worm's view, horizontal view, bird's view) already mentioned in the Codex Huygens. The third *mentita* concerns paintings in a worm's eye view, "which are placed above the line of the horizon, so that the parts behind it vanish and the parts on the foreground rise above and no limb is superimposed."[77]

---

76 I thank Prof. Pietro Roccasecca for the suggestion.

77 Ciardi, *Scritti*, II, p. 236: "Per questa veduta tutte le figure o corpi che sono sopra l'occhio si mostrano per le parti da basso, o più, o meno, secondo che sono in alto sopra la parete

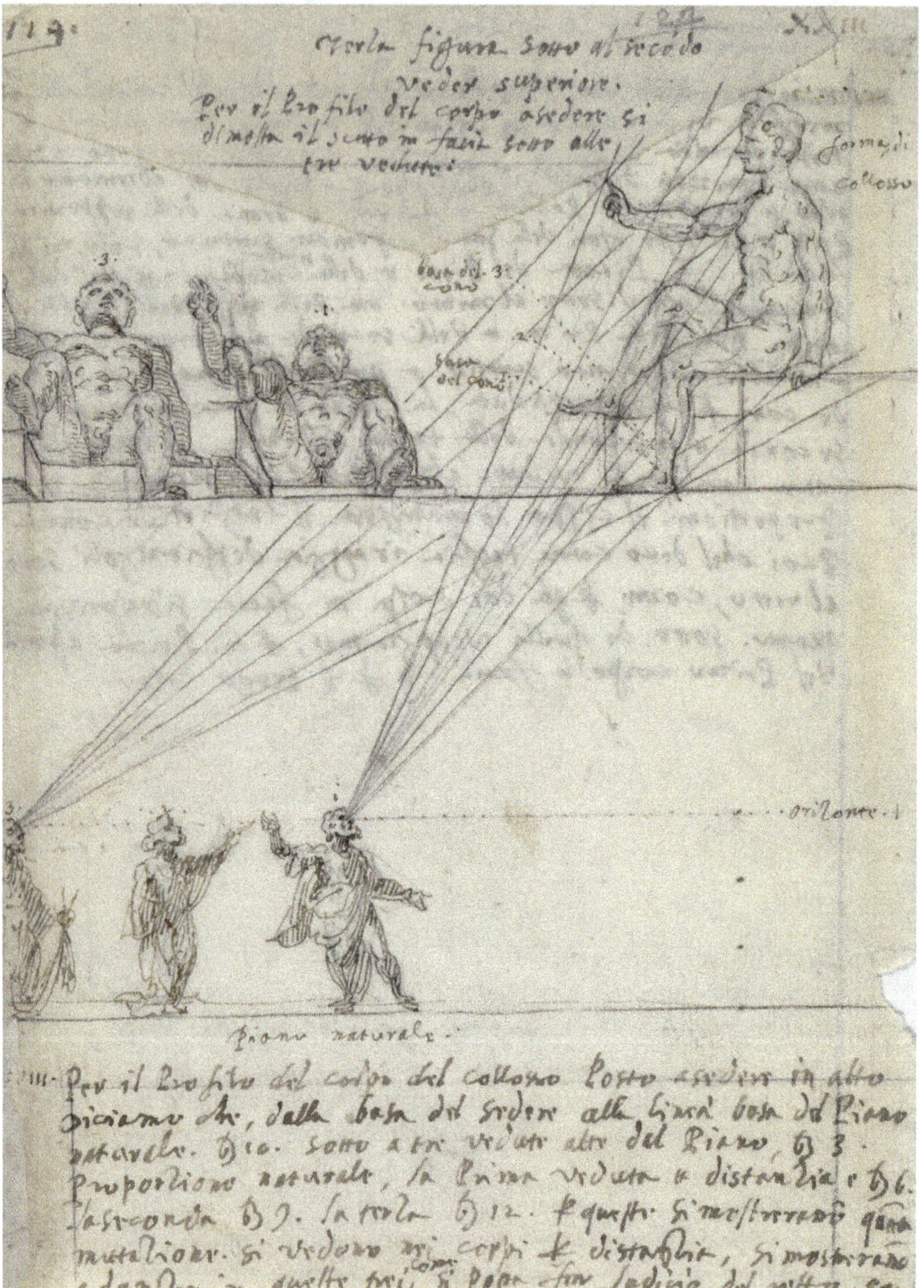

FIGURE 5.8 *Carlo Urbino,* View from below, *Codex Huygens, fol. 114, c. 1580, New York, Pierpont Morgan Library*

As one example, the author takes the series of *Triumphs of Cesar* by Andrea Mantegna (one of the governors of the *Idea*), which are made "with order and intelligence". It is not the only praise Lomazzo gives to the painter regarding his ability to use perspective: in fact, he dedicates a passage to him in his poem *Lodi d'Apelle e di altri pittori*, in which he states that Mantegna was very skilled in placing figures according to perspective: "Mantegna was as good in making figures in perspective as the wise Asclepidor/Placing the distances as if it was from life itself."[78] In another poem dedicated to the painter, he also praised his ability to paint both perspective and light, connecting again the two subjects and declaring that no painter equals Mantegna: "He discovered the numeration in perspective/And in light and in sight, which is the foundation of all our art/No one could equal Mantegna."[79]

The fourth *vista* concerns paintings seen from a normal view (mezzana), which is, according to the author, "the least foreshortened of the three", but is nonetheless difficult to obtain. Among others, he cites a Christ descending the Cross made by Bramantino above the door of the church of S. Sepolcro in Milan, which Ciardi identified as the one now in the Ambrosiana Gallery.[80]

The fifth *vista mentita* includes "all the figures which are seen from above, either much or little, under the horizon."[81] This refers to all the paintings placed on a vertical surface in bird's eye view, such as Michelangelo's *Last Judgement* or *The Crucefixion of Saint Peter* in Rome.

A discourse in itself should be done for the last *vista mentita*, "that is *intrante*, which makes us see the bodies layed on the floor, in foreshortening, which seem to enter the walls."[82] Here Lomazzo is talking about the orthogonal foreshortening, as we can find in Mantegna's lamentation in the Brera, "which is

all'orizzonte. Per il che le parti di dietro scaggiono e quelle davanti sagliono in alto e niun membro occupa l'altro." (Lomazzo, *Trattato*, p. 271).

78 Lomazzo, *Rime*, p. 91: "Al saggio Asclepidor in Prospettiva/Il Mantegna fu ugual nelle figure/Ponendo sue distanze al par del vivo."

79 *Rime*, p. 95: "Scoperse in Prospettiva l'alte scale/E nei lumi e veder dove si fonda/Tutta quest'arte, a cui ciascun s'apponda/Al Mantegna non fu pittor uguale."

80 Ciardi, *Scritti*, II, p. 237: "Questa è la manco scurzata che sia." (Lomazzo, *Trattato*, p. 271).

81 Ciardi, *Scritti*, II, p. 237: "Tutte le figure che si veggono per di sopra, o poco, o assai, sopra una faccia, cioè sotto l'orizzonte, da questa vista vengono formate" (Lomazzo, *Trattato*, p. 272).

82 Ciardi, *Scritti*, II, p. 238: "E sono quelli che paiono totalmente entrare nel muro, facendosi nel medesimo loco, per esempio, al dritto dell'occhio." (Lomazzo, *Trattato*, p. 272).

not a *vista* corresponding to the five others, but it is due to a special position of the object which may occur in any of the real *viste*" (fig. 5.9).[83]

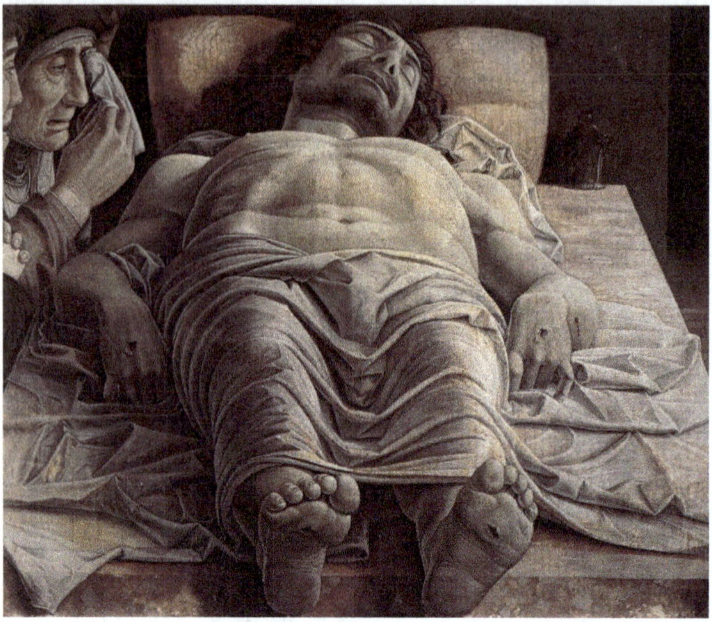

FIGURE 5.9 *Andrea Mantegna,* Lamentation of the Christ, *c. 1480, Milan, Pinacoteca di Brera*

In the *pratica,* Lomazzo will sum up these ways of sight according to the *linee rette* in ch. XIII, *Delle regole della prospettiva*:

> And finally with this art of straight lines one can conclude the discourse on the five ways of seeing. Of these five lines the first three are: one above the eye, which falls perpendicular above it as in zenith, the other that falls below it, and the third one which is in the middle (…) The other two lines are oblique, from the eye itself they come above and below and they are useful to the first three lines aforementioned. (…) So that the true eye comes to be the real mean of the true narrative, or we can say it becomes the mirror, such as ancient and modern *perspettivi* state.[84]

---

83 Panofsky, *Codex*, p. 101.

84 Ciardi, *Scritti*, II, p. 274: "Ultimamente per quest'arte delle linee rette si concludono i cinque modi del vedere. E di queste linee tre sono le prime: una è sopra a l'occhio, che gli cade perpendicolarmente sopra a guisa di zenit, l'altra che gli cade sotto, e la terza è la linea di mezzo. (…) Le altre due linee sono diagonali, che dall'occhio stesso vengono una

After this long and exhaustive categorization of foreshortenings, the author comes back to a reflection on the movements and elevation of the body (*Delle flessioni,* ch. XIX, and *Delle levazioni,* ch. XX), quoting Dürer's *De Varietate Figurarum* (1533). He underlines the necessity of understanding human movements in order to build the right perspective on each occasion, "such as Alberto showed, in different figures and heads, where clearly he shows the different quantities and the faces that look above and below."[85] With these words Lomazzo links the subject of perspective to the vastly developed subject of proportion and to anatomy, ending the chapter with the revealing words: "But I will not say more about these things, because they are more adapted to drawing than to writing." In doing so, he confirms that painters who want to learn subjects such as anatomy should look at drawings and they should practice by copying them instead of reading descriptions of the parts of the body.[86]

The last chapters of the book on perspective are perhaps the most odd, or, in other words, it seems to be difficult to find a reason for their inclusion. As was mentioned at the beginning, they are dedicated to the perspective of Bramantino. Lomazzo declares that he read many writings by him, as well as a treatise on perspective by Bernardo Zenale dedicated to his son, and one by the Milanese Vincenzo Foppa. Although the painter in this part of the book promises "to publish (*dar fuori*) this book, which is completed with pen-sketches to exemplify the writing, so that it includes almost everything that Dürer took into account in his *Simmetria*", there is unfortunately no record of this treatise, and we can legitimately doubt that it was ever published.[87] Again, it is hard to say whether the information in these books are completely reliable, but what is perhaps even more puzzling is why Lomazzo felt

---

all'alto e l'altra al basso, e servono alle tre sopradette. (…) Sì che l'occhio vero viene ad essere il mezzo della vera istoria, o vogliam dire specchio, sì come affermano gl'antichi e i moderni prospettivi." (Lomazzo, *Trattato*, p. 317).

85 Ciardi, *Scritti*, II, p. 238: "Come ha mostrato Alberto in diverse teste e figure dove chiaramente con tal ordine mostra a portar una quantità in un'altra et a formar faccie che sguardino all'insù et altre all'ingiù." (Lomazzo, *Trattato*, p. 273).

86 Ciardi, *Scritti*, II, p. 239: "Nondimeno mi risolvo di tacerle per ora, perciochè che s'aspettano più al disegno che alla scrittura." (Lomazzo, *Trattato*, p. 273). For a discussion on the issue of learning anatomy, see the next chapter.

87 Ciardi, *Scritti*, II, p. 240: "Ben prometto di dar fuori una volta certa opera vecchia di Vincenzo Foppa milanese, nella quale, oltre quello che a di lungi ne scrive, vi sono anco gli schizzi fatti con penna, sì che si comprende quasi tutto ciò che ha trattato poi in gran parte Alberto Durero nella sua Simmetria." (Lomazzo, *Trattato*, p. 275).

the need to discuss the perspective of Bramantino itself. One may suppose that he decided to include it as this type of perspective held a particular relevance, whether for his practice or for his theoretical discourse, but this does not seem to be the case. The perspective of Bramantino is divided into three parts, the first consists of using perspective from "the needs of the eye" (and again the role of experience gains central importance), the second using *fantasia*, and the third using the *graticola*, which the author does not recommend, as he is in favor of a more theoretical construction. Apart from the first *prospettiva*, which the author certainly agreed with, we can only suppose that he decided to include Bramantino's categorization of perspective both because he admired his skillfulness in using foreshortening, as the author states in different parts of his writings, and probably because of its lasting relevance in the pictorial context of the city in Lomazzo's time.[88] Bartolomeo Suardi called Bramantino (1455–1536), a renowned Milanese painter, worked for a long period in Milan, especially in cooperation with his master, the artist Donato Bramante: the latter made an impressive perspective view behind the altar of the church of Santa Maria Presso San Satiro. It was perhaps the best illusion of space in the city at the time, of which Lomazzo certainly had knowledge of. The author also praises Bramante in his *Rime*, in a poem dedicated to Raphael, mentioning the edition of the latter's writings on architecture made by Baldassarre Peruzzi: "Egli ha mostrato l'alta Prospettiva/Che da Bramante trapassò il Peruzzi/Con somma grazia posta al suo vedere."[89]

Practical Suggestions and Links with Northern Theories

At the beginning of the book on perspective the author states that "although some painters thought that Michelangelo made his foreshortenings taken from models, they are truly mistaken: he, who knew well these things, used the arts of the *flessioni* and the *trasportazioni* in making all his foreshortenings."[90]

88 Lomazzo mentions several times the *scorti* of Bramantino in different writings, especially in the *Rime* and in the *Libro de' Sogni*.

89 Lomazzo, *Rime*, II, p. 92. On Bramante and Bramantino see the recent exhibition at the Pinacoteca di Brera in Milan entitled "Bramante a Milano: Le Arti in Lombardia 1477–1499" (4th December 2014 - 22nd March 2015).

90 Ciardi, *Scritti*, II, p. 221: "E se bene ad alcuni pittori è parso che Michel Angelo facesse i suoi scorti ritirandogli da i modelli, nondimeno si gabbano e di grosso; perciò che egli, che era intelligente di queste cose, si valse dell'arte delle flessioni e trasportazioni." (Lomazzo, *Trattato*, p. 252).

Lomazzo discusses the method of transportations, called "transformations" or parallel projections in the Codex Huygens, "by means of which any required diagram can be developed from any two others, provided that the latter are located in planes perpendicular to each other."[91]

Different artists (most notably Dürer and Piero della Francesca) already used this method of transposing figures through orthographic projections and applying it to the parts of the human body. It was a method that was known and used not only by painters but also by mathematicians of the time, such as Danti and Barbaro, among others.[92] It is likely that Lomazzo drew this method from the latter's *Prospettiva Pratica* (Venice, 1569): the author quotes Barbaro more than once in the book, and his technical discourse on how to make perspective seems to have different points in common with Barbaro's practical perspective.

In particular, one can find a striking similarity with Barbaro's explanation of the perspective pyramid, on the active role of the eye, and on the quantity of distances, while the author's system of calculation seems to differ, as already noted by Panofsky, from other *perspettivi* such as Danti.[93] The latter for instance suggests a distance of at least two times as large as the maximum dimension of the panel, yet Lomazzo states instead that "the distance should be three times as large as the surface that one sees, and also in the canvas or figures one must take the distance three times as large as the height of the figure." He defined this distance as "more proportioned to the eye (…), although this decision lays in the intellect (*meaning the personal judgment*) of the practitioner, I will not make this discourse longer."[94]

Unsurprisingly, the discourse on transpositions can be found again in the contemporary Codex Huygens, especially regarding both the head and the body, though the author of the manuscript does not strictly follow the mathematical lines, especially for the parallel projection of the head.[95]

In the *Idea* he will list the elements that belong to the "perspective for painters", which is the one Lomazzo claims to be interested in, taking into account: the eye, the object, the distance and the pyramid (*Il taglio della piramide*).[96]

91 Panofsky, *Codex*, p. 103.

92 See Sven Dupré, "Galileo, Mathematical Instruments and Orthographic Projection," *Bullettin of the Scientific Instrument Society*, 2001, LXIX:10–20.

93 Panofsky, *Codex*, p. 100.

94 Ciardi, *Scritti*, II, p. 231: "Questa distanza è la più proporzionata all'occhio. (…) E perchè questa risoluzione sta nell'intelletto di colui che opera, non starò qui a renderne la ragione con lunga dichiarazione di parole." (Lomazzo, *Trattato*, p. 264).

95 See Panofky, *Codex*, pp. 30–31.

96 Lomazzo, *Idea*, ch. XXIII, in Chai, 2013, p. 107.

However, it is in the chapter, "On the eye as instrument to see the rays" that he develops this discourse, stating that:

> Furthermore the eye is called the cone of the pyramid, because all the space between the object and the lines or beams which pass along, and are extended from the extremities of the object to the point of the pyramid, end therein, as it were in the point of the cone thereof. Wherefore all the shapes and resemblances of things seen end in the eye, which together with the understanding ought to judge of their true forms, so that again it may be able to imitate them perfectly.[97]

This extended discourse is substantially similar to what Barbaro states when describing the function of the eye:

> Scholars of perspective call the eye the centre, sign and point, and it is the beginning and basis of all the expertise in perspective. Because that is the extremity and the peak of that pyramid that one must make in the way of seeing: I will now talk about how you must understand it.[98]

We can see that Lomazzo derives a theoretical way of making perspective from the geometrical construction of the pyramid, and not from the use of models or from the more straightforward method of the *griglia prospettica*. From a strictly practical point of view, this embedding of theories on perspective and optics in order to make foreshortened images from a mathematical construction may have been difficult for painters to actually adopt in their workshops.[99]

---

97 Ciardi, *Scritti*, II, p. 230: "È ancora dimandato poi l'occhio cono della piramide, perciò che tutto quello spazio che è tra oggetto e le linee, ovvero raggi estensi delle parti esteriori dell'oggetto, alla punta della piramide passa, e va a finire in esso, si come in punto over cono di essa. Per il che tutte le sembianze delle cose viste finiscono a l'occhio, sì come a quello che della cognizione, secondo le forme sue, ha da dar con lo spirito il giudizio, acciò che di nuovo ne possa partorir di simili a quello." (Lomazzo, *Trattato*, p. 262).

98 Daniele Barbaro, *Della Pratica della Perspettiva* (Venice: Appresso Camillo, & Rutilio Borgominieri fratelli, 1569), ch. II, *Dell'occhio*, p. 6: "L'occhio da perspettivi centro, segno, e punto si chiama, è il principio e il fondamento di tutta la perizia e la prova della prospettiva. Perciochè in quello è la punta, e la sommità di quella piramide che si vuol fare nel modo del vedere: il che come se ne intenda dirò brevemente."

99 See Kemp, *The Science of Art*, p. 84: "There was a growing sense, by the time that Lomazzo's books were published, that this measure of technical elaboration was missing the true point of artistic endeavor. (...) No artist or theorist seems to have gone so far as to deny

The theoretical construction of perspective in Lomazzo's book V may have been too complex and distant from the reality of the workshop. There are, nonetheless, other parts of Lomazzo's treatise in which the author gives practical suggestions on how to depict bodies according to the eye, using models and instruments such as the *telaro*. We find, for instance, detailed instructions in chapter XIV of the book *Della Pratica*; here the author links the discourse on perspective and transportations to that vastly developed subject of proportion, presenting instructions on "how to show the natural proportion according to the eye":

> First you will take a frame, wide as a sheet on a board, behind this frame you will have all the figures, round or whatever shape they are, short or tall, geometrically drawn. (…) The figure will be six *oncie* tall, you will do this in order: first from the head you will draw four equal lines, the first will depart from the top of the forehead, and there you will draw a dot. Here will start the nose, then the eyes and, needless to say, the eyelashes.[100]

Beginning from this passage, the author's discourse focuses on how to practically make a human figure with correct proportions, which can be mastered only with practice and the study of drawings (a discourse similar, as we shall see, to the parts of his writings dedicated to anatomy). Proportions and transportations are what many artists, according to the author, have written about, as he states in the chapter entitled *D'onde nascono tutte le proporzioni*:

> Now if any one may want to learn the most exact and smallest parts of these proportions, together with the way how to transfer them from one

---

that linear perspective was a necessary tool for painters, but some influential voices were being raised to the effect that perspective was not the foundation of art as it had been for some earlier theorists."

100 Ciardi, *Scritti*, II, p. 275: "Prima piglierai un telaro in piedi di larghezza di un foglio sopra una tavola, di dietro dal qual telaro si hanno a mostrare tutte le figure quadre o tonde e di qualunque altra sorte, o alte, o basse, essendo geometricamente disegnate; (…) Ora, primieramente la figura vuole esser alta sei onze; poi si ha da tenere quest'ordine: prima alla sommità del capo farai quattro linee uguali, delle quali la prima si estenda alla sommità della fronte, e quivi farai un punto alla fine della fronte, dove principia il naso; e di qua e di là segnerai gl'occhi e le ciglia, il che agl'intendenti non occorre dirlo." (Lomazzo, *Trattato*, pp. 317–318).

> body to another, I refer him to the works of Leonardo, Bramante, Vincenzo Foppa, Zenale, and for prints to Albrecht Dürer and others.[101]

This passage belongs to the first book on proportion, and for the first time in the book it introduces the reference to Dürer's prints. The latter, as different scholars noted, is an important source for Lomazzo's theory of proportion. Apart from his theoretical writings, the painter also saw many of Dürer's woodcuts (some scholars such as Rogier van Son even argue that Lomazzo was an avid collector of these prints), as references are often circumstantial when he mentions them:

> Such as Raphael, Polidoro, and Albrecht Dürer, painter, though from a barbaric *maniera*, expert and most intelligent, who made many *istorie, fantasie, guerre e capricci* which are all skillfully made, as one can see from the woodcuts that he did.[102]

However, on closer inspection we can see that Dürer is not the only inspiration for the discourse on proportion and perspective; the passage in ch. XXXI is strikingly similar to the aforementioned one in the book of practice, in which the author mentions the *quadrature* by Foppa and others:

> Now for what concerns the figures *quadrate*, Vincenzo Foppa drew many of them (...) and following him Bramante then made a book, from which Raphael, Polidoro and Gaudenzio all learned. (...) And apart from that, in many other ways one can draw a man from schemes, as for instance

---

101 Ciardi, *Scritti*, II, p. 90: "Or chi volesse intendere le minute parti delle proporzioni e trasportazioni sue da l'un corpo all'altro, vegga le opere disegnate di mano di Leonardo Vinci, di Bramante, di Vincenzo Foppa, di Bernardo Zenale; e di quelle che sono poste in stampa vegga le opere di Alberto Durero e d'altri." (Lomazzo, *Trattato*, pp. 100–101).

102 Ciardi, *Scritti*, II, p. 222: "Come a Raffaello, a Polidoro et ad Alberto Durero, pittore, benchè tenesse una maniera barbara, studiosissimo et intelligentissimo, che solo ha fatto più istorie, fantasie, guerre e capricci, che non hanno fatto, per così dire, tutti gl'altri insieme, che tutte sono ben collocate, come si vede per il gran fascio delle sue carte tagliate da lui." (Lomazzo, *Trattato*, p. 253). Other parts in which Dürer is mentioned are: ch. XIX, *Delle flessioni*, ch. XXVIII, *Della proporzione dell'ordine composito*. For the discussion on the *moti*, see ch. XVI, *Proporzioni geometriche da trasferire alla vista*, ch. XIX, *Della via di tirare i colossi alla vista*. In the *Idea:* ch. IV, *De gli scrittori dell' arte antichi e moderni*, ch. VIII, *Delle scienze necessarie al pittore*, ch. XXVI, *Del modo di conoscere e constituire le proporzioni secondo la bellezza*, and ch. XXXV, *Delle misure uguali delle membra del corpo umano e come da quelle nascono le proporzioni e le armonie.*

> thanks to the numbers of Leon Battista Alberti, with the *telaro* and the *graticola* of Albrecht Dürer and of Giovanni di Frisia di Graminge, whose instruments I have seen together with many other figures drawn by others with the perspective of Gio. Lenclaer.[103]

Here Lomazzo declares to have first-hand knowledge of Northern authors, besides Dürer, who developed methods to create perspective, such as Hans Vredeman de Vries and Hans Lencker.[104] He mentions in passing that he "saw the instruments" of these authors, and "many other drawings made with the perspective of Lencker." As discussed in part I, the argument that Lomazzo may have visited Antwerp in his youth gains substance with these words, though, as we already stated, the ideas and writings of Northern authors were circulating in Italy at the time and they were probably available in Milan too.[105]

In the *Idea,* six years later, Lomazzo mentions after the name of Dürer another "German who made the lines in perspective" which one could speculate, as Ciardi does, to be Lencker. He also adds "Isibel Peun" to the list, that is the printmaker Hans Sebald Beham, "who drew horses situated in perspective" (see fig. 5.10 and 5.11).[106] It is very possible that the author saw the works of Beham in person, as we can see this artist's influence in the pictorial work of our author, especially regarding the depiction of landscapes seen in many of Beham's works (fig. 5.12). This Northern influence on Lomazzo's pictorial work has already been noted by different scholars and it is particularly recognizable

---

103 Ciardi, *Scritti*, II, p. 277: "Or quanto alle figure quadrate ne disegnò assai Vincenzo Foppa (...) E seguendo lui ne disegnò poi Bramante un libro, da cui Raffaello, Polidoro e Gaudenzio ne cavarono grandissimo giovamento (...). Oltra di questo, in altri modi si possono crescere dalle piante i corpi umani, come per forza di numeri col velo di Leon Battista Alberto, col telaro e la graticola di Alberto Durero e di Giovanni di Frisia di Graminge, i quali istromenti io ho veduti, insieme con molte altre figure disegnate da molti, con la prospettiva di Gio. Lenclaer." (Lomazzo, *Trattato*, p. 320).

104 See the collections of drawings and schemes of Hans Vredeman de Vries, *Artis perspectivae plurium generum elegantissimae formulae antea numquam impressae* (1568), and Hans Lencker, *Perspectiva literaria* (1567), and the more textual didactic books *Perspectiva* (1571) and *Perspective* (1604).

105 Later in the book the painter also mentions a certain "Giovan Friso" goldsmith, but, as Ciardi notes, the reference is too generic and the epithet too limited to consider this reference more than a coincidence of names. See Ciardi, *Scritti*, II, p. 371.

106 Lomazzo, *Idea,* translated by Chai, p. 38.

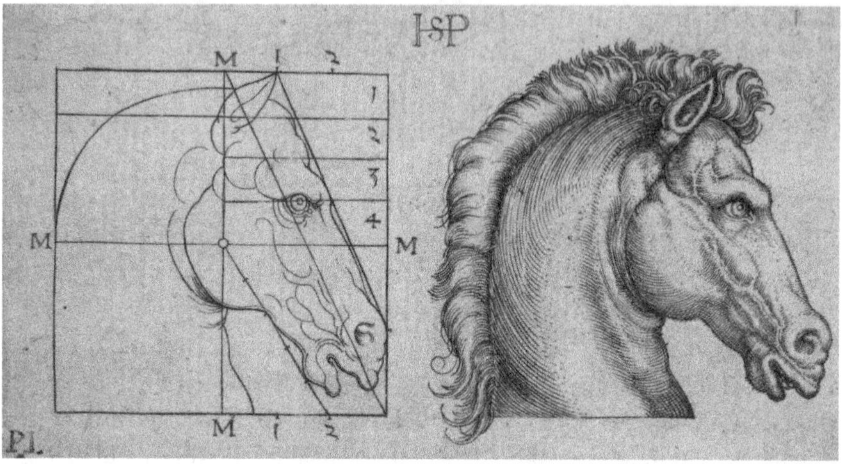

FIGURE 5.10 *Hans Sebald Beham,* Proportions of a horse head, *c. 1530, Berlin, Kupferstichkabinett*

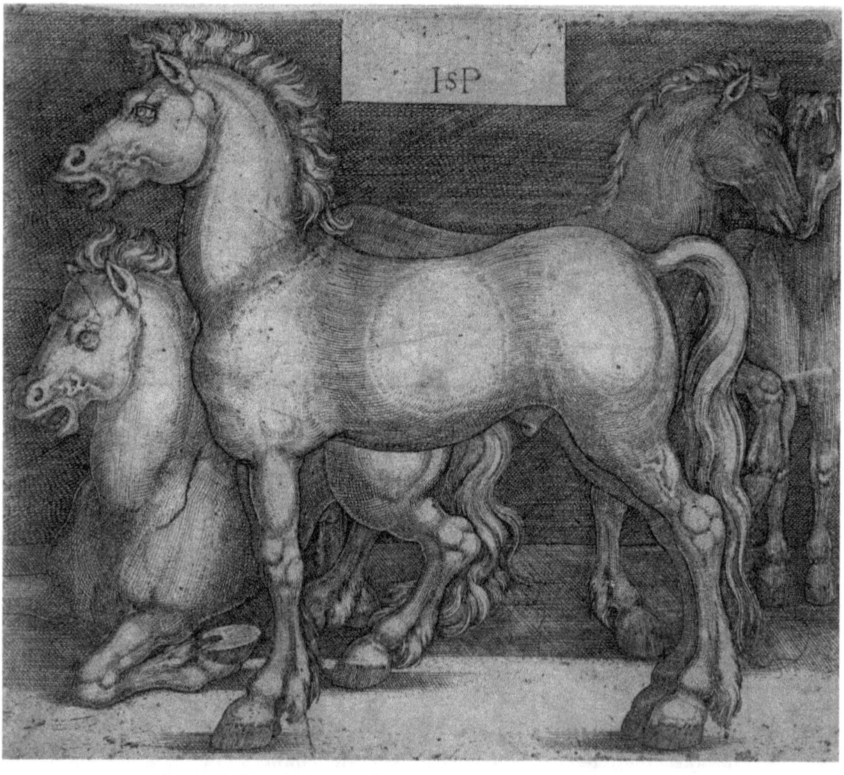

FIGURE 5.11 *Hans Sebald Beham,* Four horses, *c. 1530, Berlin, Kupferstichkabinett*

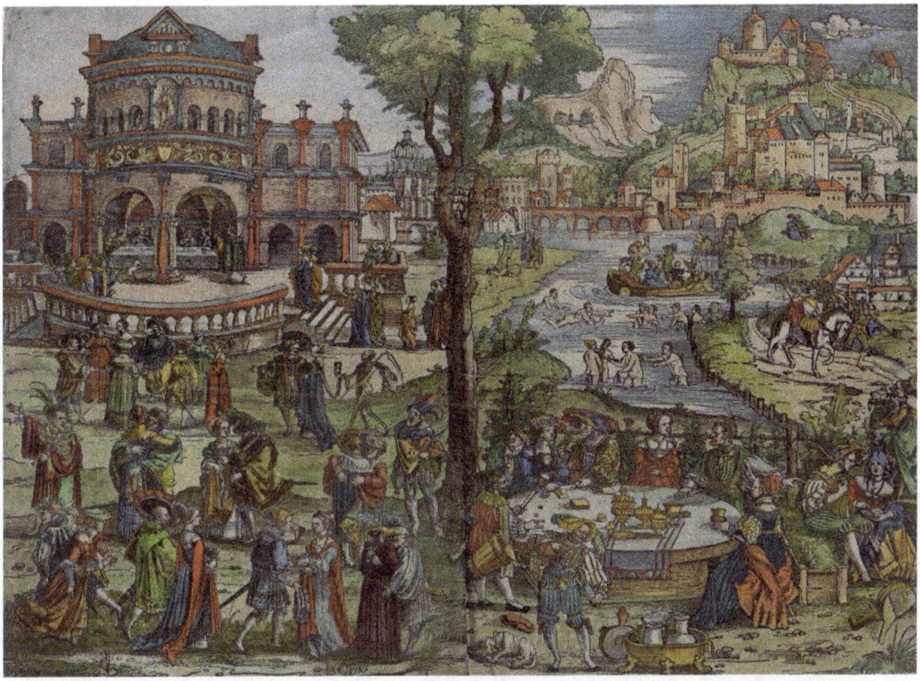

FIGURE 5.12 *Hans Sebald Beham,* The Feast of Herod, *c. 1530, Berlin, Kupferstichkabinett*

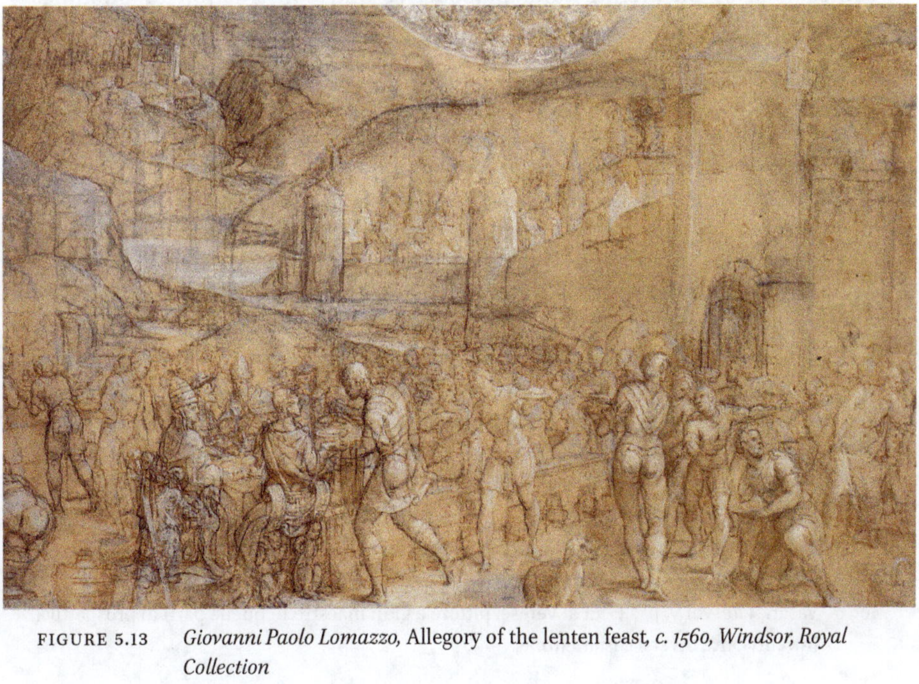

FIGURE 5.13 *Giovanni Paolo Lomazzo,* Allegory of the lenten feast, *c. 1560, Windsor, Royal Collection*

in the depiction of the landscape of the now lost *Cena Quadragesimale* in Piacenza (fig. 5.13).[107]

The references to Northern "scholars of perspective", though scarce, denote a clear interest by the author not only for the theory but also for the practice of making perspective that these artists were using.

In the theoretical works of these authors, one cannot but notice different similarities between the didactic intent explicated in Lomazzo and works such as Vredeman de Vries' *Perspective*, a book dedicated to artists that aimed to instruct painters as well as "all lovers of the subject" on how to make images with perspective. De Vries explained architectural and mathematical rules with images rather than words in order to illustrate proper foreshortening techniques. As in Lomazzo's book on perspective, practical suggestions in de Vries' book are often embedded in a description of previous theories, such as Vitruvius' theory of five orders: "We have offered you, dear lovers of the Art of Perspective, the foregoing as we thought most fit, so that you could derive satisfaction from its theory and principle, and so as to make you enjoy it even more, we have added the five ancient orders, according to the theory of the Roman architect Vitruvius."[108]

Vredeman de Vries, who was already defined by Vasari in 1568 as "painter and great master in those parts of perspective", included many projections employing one-and multi-point perspective in his book.[109] His practical discourse focuses on architectural perspective and construction of buildings

107 On the subject, see Maria Virginia Cardi, "Gian Paolo Lomazzo: La Cena Quadragesimale," *Bollettino Storico Piacentino*, 1996, *I*: 77–89.

108 Hans Vredeman de Vries, *Perspective, Das ist Die weitberuembte khunst, eines scheinenden in oder durchsehenden augengesichts Puncten, auff vnd an eben stehender Wandt und Mauren, Taffelen oder gespannenen Tuech: in welchem anzuschewen sien mögen die gebewde der Kirchen, Tempeln ... auff die alte vnd newe manier, vnd mehe dergleichen gestaltnußen alhie furgestelt, alles auff seine eigene fundamental Linien, vnd das fundament der selben eigentlich außgelegt mit dere selben artlichen beschreibung; Allen Mahlern, Kupfferstechern .... zu ihrem Stvdiren sehr angenehm lieblich vnd nutzbar* (Leiden: Henricus Hondius, 1604), XLVIII–XLIX: "Omdat u, Liefhebbers van de perspectiefkunst genoegen mag beleven aan de leer en het beginsel, hebben wij het voorafgaande naar ons beste goeddunken aangeboden en opdat het u nog meer plezier zou doen hebben wij hier nog de vijf antieke kolommen naar de leer van de romeinse architec Vitruvius bijgevoegd." Translated in *Perspective*, ed. by Peter Karstkarel, Tableau B. V., 1979, p. 38. On the subject De Vries already published years before the *Artis Perspectivae Plurium Generum Elegantissimae Formulae* (Antwerp: Gerard de Jode, 1568.)

109 Vasari, *Vite*, vol V, p. 24: "Et al Verese, pittore e gran maestro in quelle parti di prospettiva, in venti carte diversi casamenti."

according to the right proportions; the foreshortening of human figures, similar to techniques explained by Lomazzo, are also always placed in the context of architectural spaces. The *Perspectiva* by the goldsmith from Nuremberg Hans Lencker, published in 1571, also teaches the doctrine of perspective by means of illustrations (eleven woodcuts of geometrical figures), which is "a noble art known to physicians and other authorities on nature and the heavens."[110] The extent to which Lomazzo knew these authors' writings cannot be pinned down with certainty, but there is little doubt that he had knowledge of their work as *prospettivi* and how they linked the rules of perspective with the theory of architecture in their work, which may have inspired him to include the discourse on architecture we have seen in the previous sections.

Lomazzo was certainly not alone in his knowledge of these authors; perhaps the most direct link is the introduction to Vignola's *prospettiva* by Danti, who also refers to Northern scholars of perspective when giving a brief history of this subject:

> We also have abbreviated versions of these common rules, given by Leon Battista Alberti, by Leonardo da Vinci, by Alberto Duro, Giovacchino Fortio and Giovan Lencker and Wenceslao Giannizzero of Nurenberg, who has given perspective renderings of the regular bodies and other, composite bodies as was done by Pietro dal Borgo, although Fra Luca later printed them under his own name.[111]

Here Danti lists the Dutch Joachim Fortius Ringelbergius, that is the astrologer and mathematician Joachim Sterck van Ringelbergh who published the *Liber de Ratione Studii* (1541) with a section on optics, including both natural and artificial perspective.[112] More interestingly, he quotes Johannes Lencker along

110 Hans Lencker, *Perspectiva* (Nuremberg 1571), AIII, V: "Und wiewohl die edle perspectiva dermassen ein hoe/schöne/suptile/ (jedoch weitleufftige) tunstist/ wie dann das den Phisicis und naurtundigern des gleichen des Gestirn erfahrnen wolbetande welche sich desselben auch bis zu den himlichen Corperen zugebrachten wissen." See Christopher S. Wood, "The Perspective Treatise in Ruins: Lorenz Stoer, *Geometria et Perspectiva*, 1567," *Studies in the History of Art*, Symposium Papers XXXVI: The Treatise on Perspective: Published and Unpublished, 2003, *LIX*:234–257, p. 236. On the education of artists in Germany see for instance L. Schmitt, "Education and Learning among Sixteenth-Century German Artists," *Studies in the History of Art*, Symposium Papers XXXVII: Hans Holbein: Paintings, Prints, and Reception, 2001, *LX*:72–81.

111 Iacopo Barozzi da Vignola, *Le Due Regole*, translated in Field, "Giovanni Battista Benedetti", p. 72.

112 See Joachim Sterck Van Ringelbergh (1499–1536), *Liber de Ratione Studii* (Basel: Bartholomeus Westheimer, 1541).

with another goldsmith, Wentzel Jamnitzer (1508–1585), stating later in the book that he saw a copy of Jamnitzer's *Perspectiva Corporum Regularium* (Nuremberg 1568). This is another indication that writings of these Northern authors were probably available in Italy at the time.

It is reasonable to doubt whether Lomazzo took his reference to Lencker from Danti's list of authors, though it is peculiar of him to state that he had personally seen the instruments of Lencker and De Vries. Regardless of the reliability of this last statement, the author certainly shows a particular interest for the workshop practices of these artists from the North, which he wants to transmit in his writings and this constitutes a rarity in Italian treatises on art of the time.

Conclusions

Lomazzo's treatment of perspective is a mixture of previous discourses by more traditional authors on art (Serlio, Alberti, Leonardo) and optical contemporary theories (Vignola, Barbaro, Benedetti). The author embeds practical suggestions for painters on "how to make the right perspective" into an exhaustive theoretical discussion on the role and the function of the eye. Moreover, he includes an extended discourse on architectural issues presenting different practical examples and methods of foreshortening for artists to use. Though this book has always been considered the most complicated of all five books and it was often criticized as not being consistent in all its parts, I showed how this book is ultimately a product of the discourse present in Milan at the time and of the connections the author had with different personalities of the city. Starting from the concepts that Lomazzo presents, I attempted to pin down the context in which this book was written, in order to understand the material that he gathers, as I have done for the chapter on color. Moreover, I showed how the notions the painter conveyed in this book are closely connected with other parts of his writings. In fact, despite the author's strict organization and categorization of the different parts of perspective, to gain an overall understanding of the subject he discusses one cannot separate *prospettiva* from the discourse on the power of the eye presented in the book of light or from the practical suggestions offered in the book of practice. Despite Lomazzo's references mostly to ancient sources (Aristotle or Euclid), it is clear that the author also borrows from optical theories presented by contemporary authors. Finally, I enquired on Lomazzo's connections with contemporary practitioners from outside of Italy, hinting at his link to the north of Europe, in order to make sense of the notions presented in the book and the impact that they might have had on the author's achievements. In the following chapter I will take into consideration the context of Milan to explain another fundamental topic in Lomazzo's Treatise, that is the study of the human body and painters' learning of anatomy in the city towards the end of the sixteenth century.

CHAPTER 6

The Study of the Body

The Anatomical Discourse in Lomazzo's Writings and Painters' Learning of Anatomy in Milan

To possess skills in depicting the human body was one of the basic requirements a painter in sixteenth-century Italy had to have in order to make a lifelike depiction of people in his pictorial work. Renaissance *studio* held the human figure as central for the artist's training, and anatomical images –especially drawings- were systematically produced by and for artists to understand and to faithfully reproduce the superficial muscles as well as all parts of the body.[1] Mastering anatomy was essential for artists, and in order to learn it, painters and sculptors (first privately in the workshop, later in academies of art) fostered the production of anatomical illustrations, drawings and studies that circulated in Italy from workshop to workshop.[2]

This chapter focuses on Lomazzo's discourse on anatomy, and especially on the practical suggestions that he gives to painters in order to learn how to draw a body, starting from the heritage of Leonardo and then moving to the author's categorization of the different parts of the human figure. It outlines his connections with anatomists in Milan, comparing the words that he uses in his book with the other Northern Italian source we hinted at in the previous chapter, the Codex Huygens, with the aim of understanding the author's systematization of anatomical information. Finally, it presents a series of anatomical drawings attributed to a fellow artist and member of the Blenio Academy, Annibale Fontana, to see how the circulation of anatomical illustrations derived from Leonardo gradually became wider, as they started to be seen not only in the workshops but in the environment of the *Accademia di Blenio*.

1 For a survey on the development of anatomical drawing books of anatomy see among others Mimi Cazort, Monique Kornell and Kenneth B. Roberts (eds.), *The Ingenious Machine of Nature: Four Centuries of Art and Anatomy* (Ottawa: National Gallery of Canada, 1996), especially Monique Kornell, "The Human Machine", pp. 43–70.

2 Possibly the most famous sixteenth-century example in Italy are the Carracci painters in Bologna, who taught anatomy and fostered the production of anatomical images of which we still have different examples. The Carracci stressed the importance of drawing from life. Agostino produced drawings systematically studying the parts of the human body which were later engraved and became important teaching tools for the years to come. On Italian artists' interest in anatomy, see among the latest studies Domenico Laurenza, *Artists and Anatomy in Renaissance Italy* (New York: The Metropolitan Museum of Art, 2012).

 | DOI 10.1163/9789004330269_009

In Milan, at the time Lomazzo published his treatise, artists were still under the influence of Leonardo's lesson, as he had notably been, like Michelangelo in Rome, one of the prominent innovators concerning the depiction of the human body.[3]

Speaking about the human bone-structure, Lomazzo declares in book VII of the Treatise:

> Bones are the principal part of the body. They support the body and they are the steady chain of the limbs. So it is necessary for us to see how they are put together (…) following those precepts that Leonardo wrote in his Anatomy of the Human Body, where, talking about the connections of the bones, he states that it is not possible for the painter to depict a body without knowing how the bones are shaped underneath.[4]

Certainly in Lomazzo's circle, Leonardo was still regarded as the master and the model to follow in the subject of anatomy, and it is not surprising to find a poem declaring him to be the "one from whom everybody learns." In *Di Leonardo da Vinci*, Lomazzo states: "He was able to rejoin anatomy with bright light, so that it gives the outlines to the bones and muscles, and such a depth that everyone is amazed and learns from this."[5]

Although the aim of the chapter is not to focus on Leonardo's interest and achievements regarding the anatomical representation of the body (which have been studied in some detail), it is nonetheless essential to give a short overview of the Florentine artist's interest in human anatomy and his production of a compedium of anatomical images, in order to subsequently link the

---

3 On the most recent studies on Leonardo and his anatomical drawings, among the vast literature, see Claire Joan Farago, *The Fabrication of Leonardo da Vinci's* Trattato della Pittura, *1651* (Leiden: EJ Brill, forthcoming c. 2016/7). I thank Professor Farago very much for sharing her work with me. On the influence of Leonardo's drawings in Milan see also Martin Kemp, "Drawing the Boundaries", and Pietro Marani, "Leonard's Drawings in Milan and their Influence on the Graphic Work of Milanese Artists", both in *Leonardo da Vinci Master Draftsman*, ed. by Carmen C. Bambach (New York: The Metropolitan Museum of Art, 2003).

4 Ciardi, *Scritti*, II, p. 535: "Non è chi non sappi che principale parte sono le ossa. Conciosiache sono il sostegno e termini delle membra e la vera salda catena loro. Onde è necessario che vediamo in qual modo fra loro si compongano (…), secondo quel precetto che già Leonardo lasciò scritto nella sua anatomia del corpo umano, là dove, parlando de l'ossa et incatenatura loro, dice di non essere possibile che il pittore faccia con ragione un corpo senza sapere come stiano l'ossa principalmente sotto." (*Trattato*, p. 615–616).

5 Lomazzo, *Rime*, p. 94: "Seppe congiunger sì coi chiari lumi/L'Anatomia che pare ch'ella stessa/L'ossa e i muscol contorni, e ch'il rilievo/Gli dii cotal, ch'ognun stupito parte."

latter with the discourse that Lomazzo presents in his treatise as well as with the achievements of the artists in his circle.[6] As is well known, Vasari wrote about Leonardo's intention to publish a collection of anatomical images in collaboration with Marco Antonio Della Torre, a skillful anatomist, a project that should have resulted in a series of volumes, a sort of "visual anatomical encyclopedia" which is now collected in Windsor under the name of *Quaderni d'Anatomia*:

> He (*Leonardo*) afterwards gave his attention, and with increased earnestness, to the anatomy of the human frame, a study wherein Messer Marcantonio Della Torre, an eminent philosopher and himself mutually assisted and encouraged each other.
>
> Messer Marcantonio was at that time holding lectures in Pavia, and wrote on the same subject. He was one of the first, as I have heard say, who began to apply the doctrines of Galen to the elucidation of medical science, and to diffuse light over the science of anatomy, which up to that time, had been hidden in the almost total darkness of ignorance. In this attempt, Marcantonio was wonderfully aided by the genius and labour of Leonardo, who filled a book with drawings in red crayon, outlined with the pen, all copies made with the utmost care from the bodies dissected by his own hand.[7]

Leonardo not only drew anatomical images but he was believed, in Vasari's words, to have "made many anatomies (that is, dissections) performed in the

6 One footnote cannot give justice to the vast literature on Leonardo, however, for reference purpose, I quote here the studies of James S. Ackerman, "Leonardo's Eye", *Journal of the Warburg and Courtauld Institutes*, 1978, *XLI*:108–146; Martin Kemp, *Leonardo da Vinci. Le Mirabili Operazioni della Natura e dell'Uomo* (Milan: Mondadori, 1982); Kenneth Clark and Carlo Pedretti, *Leonardo da Vinci Master Draftsman* (New York: The Metropolitan Museum of Art, 1983); Ernst H. Gombrich, *Antichi Maestri, Nuove Letture: Studi sull'Arte del Rinascimento* (Turin: Einaudi 1987); Frank Fehrenbach, *Licht und Wasser: zur Dynamik naturphilosophischer Leitbilder im Werk Leonardo da Vincis* (Tuebingen: Wasmuth, 1997); Bernard Schultz, *Art and Anatomy in Renaissance Italy* (Ann Arbor (Mich.): UMI Research Press, 1985), especially ch. IV, "Leonardo and Michelangelo," pp. 67–111.

7 Vasari, *Vite*, translated by J.T. Goodrich, "Andreas Vesalius and Anatomy: a Re-evaluation of his Efforts", *Histoire des Sciences Médicales*, 1982, *17*:13–16, p. 14 ff. on the issue of the intention of publishing Leonardo's research see Farago, *The Fabrication of Leonardo da Vinci* Trattato della Pittura, especially pp. 9–16.

ospedale of Santa Maria Nuova."[8] The artist himself declares, at the end of his life, that he dissected "more than thirty bodies of men and women of all ages."[9]

As it has been recently noticed, Leonardo gave his research on the anatomical body the overall title *De Figura Humana*, a meaningful choice, as Laurenza points out, that underlines the artist's intention to take into consideration the objective form of the body, "as it appears directly to the eye."[10] This research into the causes for the external appearances of the body led him into a systematic search for and study of the single parts of the human figure through the extensiveness and accuracy of his drawings, starting from the bones and the muscles, looking for "the origin of each limb" and giving shape to what he defines as "the natural man" (*omo naturale*):

> This illustration of the human body will show you the man just as if you had the natural man in front of you. The reason is that if you want to know the parts of the anatomical man you turn it, either him or your eye, in order to see the different angles, considering it from above and below and from the sides, turning it around and looking for the origin of each limb. (…) So thanks to my drawings you will know every part and every whole.[11]

More than half a century later, artists in Milan followed the anatomical canon introduced by Leonardo, and his anatomical drawings circulated and were copied in different parts of the city.

Lomazzo does not include *anatomia* itself among the five principal parts of painting (proportion, movement, color, light, perspective) in his *Trattato dell'Arte*: he talks about the anatomy of the human figure (that is, a description of all the bones and muscles which constitutes the body) in a scattered way, mostly in connection with the much more developed subject of proportion to which the first book of the entire work is dedicated. However, anatomy is

8 Vasari, *Vite*, p. 563.

9 Jearn Paul Richter (ed. by), *The Literary Works of Leonardo da Vinci* (London: Oxford University Press, 1939), 2 vols., vol. 2, p. 393.

10 See Domenico Laurenza, *De Figura Umana* (Florence: Olschki, 2000).

11 Leonardo, c. 1508–1510, W. 19061r (K/P): "Questa mia figurazione del corpo umano ti sarà dimostra non altrementi che se tu avessi l'omo naturale innanzi. E la ragion è che se tu vuoi bene conoscere le parti dell'omo natomizzato tu lo volti, o lui o l'occhio tuo, per diverso aspetto, quello considerando di sotto e di sopra e dalli lati voltandolo e cercando l'origine di ciascun membro. (…) Adunque per il mio disegno ti sia noto ogni parte e ogni tutto."

declared in the later *Idea del Tempio della Pittura* (1590) to be "the most important and necessary discipline for drawing." As Lomazzo states:

> Yet above all, the most important and necessary discipline for drawing is anatomy. It teaches us to attach members, veins and bones, as well as to connect tendons in the bodies and construct muscles, according to the most reliable means available, by taking examples from the dead and the living bodies.[12]

He implies that the knowledge of anatomy is the basis for the preparatory drawings of figures, without which not much can be achieved in painting.

The author consequently censures those who are ignorant of this discipline, declaring that:

> Whatever expertise and practice artists might have concerning the rest, they can never approach or rival nature, they can barely imitate it, for they ignore the composition and arrangement of limbs and other parts concealed under the skin from which movement and its many and varied effects have their origins.[13]

The practice of "making anatomies" is also mentioned in book VII, when Lomazzo talks about the *modelli necessari in pittura* ("models necessary for painting"): "One must be sufficiently practiced so as to fabricate clay or wax models where the proper placement of shadows and lights on the bodies can be more easily recognized, as the most excellent artists in this have done."[14] Subsequently the discourse switches to the theory of Dürer and his division of the body, and the symmetry derived from geometry.

The subject is further developed in chapter XXV, "On the last part of painting and its species", the "last part" being form and anatomy being listed as one of its species: "Anatomical form is that which in the human body, or any other body, consists of its members, bones, and all that is necessary to make it perfect."[15]

In Lomazzo's *Trattato*, anatomy is mentioned several times, obviously in the first book on proportion, and in the last book, *Della Pratica della Pittura*. A descriptive treatment of the bodily parts is limited to chapter V of the first

---

12 Lomazzo, *Idea*, ch. VIII; translated by Chai, 2013, p. 73.

13 Ibidem.

14 Lomazzo, *Idea*; translated by Chai 2013, p. 72.

15 Lomazzo, *Idea*; translated by Chai 2013, p. 111.

book, *Dei membri esteriori del corpo umano*, where the author probably takes inspiration from Vesalius's *De Humani Corporis Fabrica* (published in 1543).[16] According to the author, this short chapter on the parts of the body should be preparatory to the study of human proportions:

> In order to be clearer, I will here mention all the external bodily parts that form the human body, since those are more necessary to the painter than the rest, in order to understand the proportions that I will treat in the subsequent chapters.[17]

For a description of the different parts of the body the author uses terms with a strong dialect flavor, adding then synonyms in *volgare*, such as he does for *zucca*, "that is sometimes called *capo* or *testa*, the upper part of the body", the chin "*mento*" or "*barbozzo*", and others (*zazzare, torti, cuticagna, grasella, mirenga, groppo, pettinicchio*). This long list shows a terminology constituted mostly by a slang of Northern flavor, such as the one that was most likely used in the painter's workshop. These words can be legitimately regarded as a reflection of those workshop practices in which a pupil had to distinguish among different bodily parts in order to understand to which one the master was referring. On this note, Lomazzo also often provides synonyms for the same part of the body, as to allow the majority of readers, regardless of their social standing, to understand.[18]

Apart from this short description, the bodily parts are rarely mentioned for their own sake, without linking them to the measures and the connections of each part which constitutes the book of proportion. The lack of a systematic discourse on the description of the human body in the first five books is then remedied in the last book on practice, where anatomy is finally given a sort of systematized discourse in two chapters, ch. XXIII (*Della forma del corpo umano e dei suoi artefici*) and ch. XXIV (*Della forma delle ossa nel corpo umano*). While in the latter the author's discourse on bones is very specific, and again recalls Vesalius (whom here the author mentions directly: "the limbs linked in the way that Vesalius suggests"), in *Della forma del corpo umano* the discourse appears more theoretical.[19] Lomazzo explains in a succinct way the development of

16 See Ciardi, *Scritti*, II, p. 41.

17 Ciardi, *Scritti*, II, p. 41: "Per maggior chiarezza verrò in questo loco a nominare tutti i membri, overo parti esteriori, che formano il corpo umano, per esser loro di più necessità che il resto al pittore per intendere le proporzioni sue che ne i seguenti capitoli si tratteranno." (Lomazzo, *Trattato*, p. 36).

18 See the following subchapter, "The Anatomical Terminology."

19 Ciardi, *Scritti*, II, p. 536: "Collegati nel modo che mostra Vesalio nei lombi." (Lomazzo, *Trattato*, p. 617).

anatomy from the ancient Greeks to Leonardo and Michelangelo, and he mentions the artists whom he considered to be the best ones in mastering this skill, among which we find Baccio Bandinelli, Aurelio Luini and Perin del Vaga, artists who will also be quoted in the *Rime*. Although the reasons for this selection are not clear in all points – apart from Luini, fellow member of the Blenio Academy – it is possible to infer that Lomazzo truly believed these artists to be very skilled in depicting the human body. The sculptor Bandinelli for instance used to make detailed preparatory drawings for his sculptures, and he copied the muscles of the human body both from other statues and from live models.

The painter Aurelio Luini (fig. 6.1), whom, as we have seen, Lomazzo knew personally since he was member of the *Accademia di Blenio* under the name of *Compà Lovign*, also used to draw studies of figures in different positions and sizes, as Perin del Vaga did.[20]

Significantly, Lomazzo gives esthetic judgments on painters' abilities to "make anatomies", which is to make anatomical drawings. We have different "critical poems" in the author's *Rime* (1587) regarding a variety of painters, (among which the name of Perin del Vaga is mentioned again) in which these painters are judged and praised on the basis of their abilities in mastering this skill:

> On Perin del Vaga: With him you see the anatomy rejoined/And skillfully expressed in such a masterly manner/That no one could surpass him later.[21]
>
> On Federico Barozzi: He reached the highest point in Anatomy/Showing the living alive, the dead as dead/And he was a divine painter in all parts.[22]
>
> On Aurelio Luini: With rare Anatomy and a profound style/He showed lights, shadows and reflections/Touching the very core of the art with the learned brush.[23]

---

20 On Luini, apart from the recent exhibition catalogue *Bernardino Luini e i suoi Figli*, see Francesco Debolini, "La Pittura Manierista di Aurelio Luini," dissertation at the University of Siena, 1995/96. On Perin del Vaga see the study of Elena Parma Armani, *Perin del Vaga: L'Anello Mancante* (Genua: Sagep, 1986).

21 Lomazzo, *Rime*, p. 98: "Di Perin del Vaga/Con ciò l'anatomia vedi congiunta/espressa con tanta arte e magistero/che più di lui non v'è ch'il pregio n'abbi."

22 Lomazzo, *Rime*, p. 102: "Di Federico Barozzi/D'Anatomia giunse al più alto segno/Morti i morti mostrando, vivi i vivi/E in ogni parte fu divin pittore."

23 Lomazzo, *Rime*, p. 106: "Con rara anatomia, con stil profondo/Di lumi, d'ombre e di riflessi, il fondo/De l'arte col pennel dotto toccando."

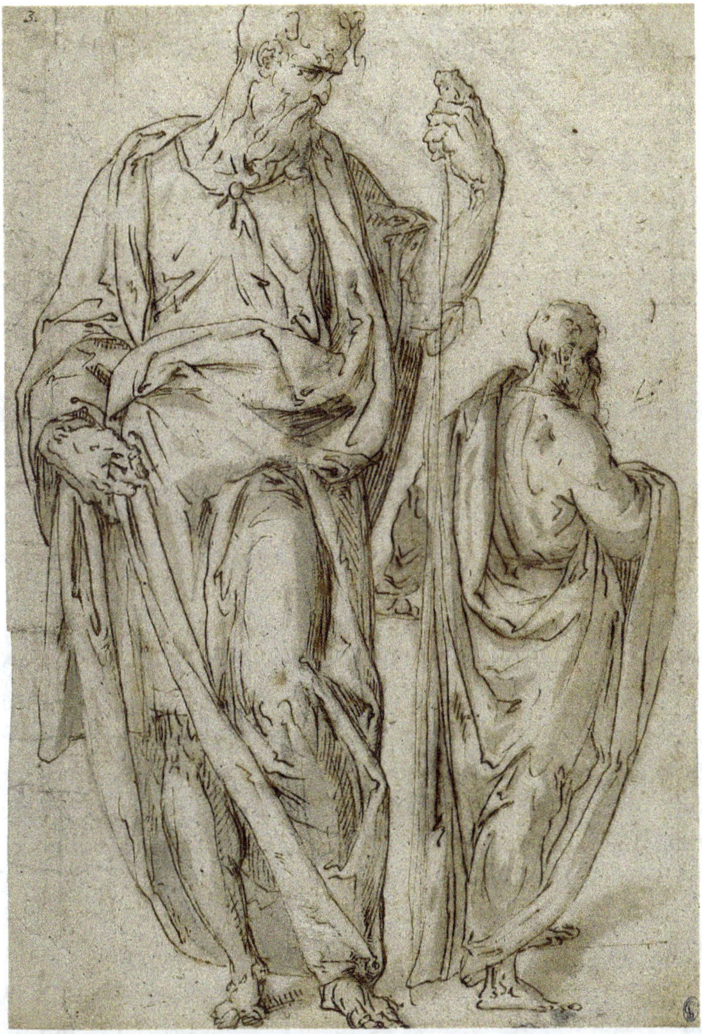

FIGURE 6.1 *Aurelio Luini,* Two standing figures, one holding a staff in his left hand, *c. 1560, Sotheby's, New York (2008)*

Moreover, the author, as we have seen in other parts of his writings, is very keen on judging "modern painters", accusing them of incompetence and of being capable of making only unfitting anatomies:

> But they (*modern painters*) make anatomies wrong (*natomie sbandite*)/ Even though they have the right light or focal point/And they use *invenzioni* which are found without much thinking/Even if they are colored/

> Such as an incompetent artist would do/Who is nonetheless all-content of his money.[24]

Lomazzo is here implying that in his own time there are many artists who care only about money and are actually unable to paint and draw properly; very possibly, he is again referring to artists such as Bernardino Campi and other "foreigners."[25]

Anatomy in Milan: Lomazzo's Connections

Anatomy, along with anatomical representations, was a subject that held a central role in the Counter Reformation discussion that was taking place in Northern Italy at that specific historical time. In painting, a naturalistic representation of the human figure was demanded as a response to protestant iconoclasm, and this necessity was in all probability more widespread in cities like Milan, which were geographically at much greater risk of falling under the Protestants' influence. Of course, stating that the Counter Reformation directly fostered anatomical studies in Milan (as in other Italian cities) would be both generic and misleading. As Cunningham pointed out, the main issue is to assess what the Counter Reformation and its defining characteristics actually were, as "the term does not embrace a readily specified series of actions or set of attitudes within the sixteenth-century Catholic church and which were identifiable at the time as related by their coherence and interconnection."[26]

Besides the terminological issue, one can legitimately infer that the study of anatomy and the practice of writing on it was, at different levels, encouraged by the Council of Trent (1545–1563), whose precepts shaped many decisions of one of its prominent participants: the cardinal Carlo Borromeo. As we saw in part I, Borromeo was the spiritual guide of Milan in Lomazzo's time, having been elected archdeacon of the city in 1566, and to him were dedicated different anatomical treatises, such as Bartolomeo Eustachi's *De Renibus* (1563) on the discovery of the adrenal glands. He also possessed many books on this subject in his own library, such as Realdo Colombo's *De Re Anatomica* (Venice

24 Lomazzo, *Rime*, p. 123: "Ma da lor son le Natomie sbandite/O ch'abbino il ver lume, o pur di foco/E l'invenzioni ritrovate a grillo/ Benchè son vagamente colorite/Come convien a chi nel arte è fioco/Ma i scudi suoi lo tengono tranquillo."

25 For the artistic climate in the city at the time see part I: "Milan and the Arts: Spaces of Interaction, Spaces of Production."

26 Andrew Cunningham, *The Anatomical Renaissance* (Farnham: Ashgate, 1997), p. 256.

1559), the already mentioned *De Humanis Corporis Fabrica* by Andrea Vesalius, (Basel 1543) and the *Opuscula Anatomica* (Venice 1571), by the anatomist Eustachi, whom Borromeo probably knew personally.[27]

In mid-sixteenth century Milan we find the presence of two illustrious anatomists: Pietro Martire Carcano and his younger brother Giovanni Battista. Pietro, who was a pupil of Vesalius, introduced in Milan a reform on the study of anatomy and taught it to his brother, Giovan Battista.[28] The latter was appointed by the Milanese *senato* to be professor of anatomy in Pavia (1573), after Gabriele Cuneo. At almost the same time, he published two treatises on the heart and on the eyes, *De Cordis Vasorum in Foetu Unione,* and *De Musculis Palpebrarum atque Oculorum Motibus Deservientibus,* which were collected and published under the title of *Anatomici Libri Duo* (1574).

In *De Vulneribus Capitis Liber Absolutissimus Triplici Sermone Contentus* (1584), Giovan Battista states how his brother and teacher Pietro Martire became a fundamental figure in the history of anatomical studies before him in the city of Milan:

> From an early age my brother Pietro Martire was considered first among anatomists and surgeons (because he started the study of anatomy in Milan from the famous school of the divine Vesalius, and in his light he started his activity) and I followed his footsteps in Milan.[29]

One year later Giovan Battista also published a book, the *Exenteratio Cadaveris Illustrissimi Cardinalis Borromaei* (1584), in which he describes the autopsy of Carlo Borromeo that he performed under the supervision of different members of the cardinal's household, a description that is one of the most interesting documents regarding contemporary post-morten examinations.[30]

---

27 *San Carlo e il suo Tempo*, pp. 173–174. On the anatomical treatises owned by Carlo Borromeo, see Agostino Saba, "La Biblioteca di San Carlo", *Fontes Ambrosiani*, Florence 1936, *12*:13–20.

28 See *San Carlo e il suo Tempo*, p. 174.

29 Giovan Battista Carcano, *De Vulneribus Capitis Liber Absolutissimus, Triplici Sermone Contentus* (Milan: apud Pacifico Da Ponte impensis Pietro Tini, 1583), folio 116v: "Quando a teneris unguiculis semper fratris mei Petri Martiris Carcani inter chirurgos, et anatomicos facile primarii (is siquidem ex celeberrima illa Divini Vesalij schola originem traxit Mediolanique primus artem anatomicam exercere coepit eamque in lucem prodire coegit) vestigia Mediolani sim insequutus."

30 Giovan Battista Carcano, *Exenteratio Cadaveris Illustrissimi Cardinalis Borrhomaei Mediolani Archiepiscopi, a Io. Baptista Carcano Leone, Anatomen in Gymnasio Ticiniensi Publice Profitente Factae, Verissima Atque Elegantissima Enarratio* (Milan: Pietro Tini,

Paolo Morigia in *La Nobiltà di Milano* (1595) states that "Carcano did such good to our city that he is worth endless praise."[31] Later, in the *Ateneo dei Letterati Milanesi* (1670) Filippo Picinelli will add that: "There was no artery, vein, nerve or muscle that he did not know distinctly, and he made everybody understand what nature is able to operate with its marvelous organization."[32] Lomazzo probably knew Giovan Battista and Pietro Martire personally, since in his autobiography at the end of the *Rime* (1587) he states that he painted portraits of both: "I made portraits of the Carcano brothers, Pietro Martire and his brother/Both great anatomists and surgeons."[33]

Along with these names the artist mentions other *dottori* whom he knew and portrayed, among them Francesco and Marco Antonio di Bossi, Alessandro Archinto, Federico Quinzio, Girolamo Dugnan and Crespo, who are all mentioned in the *Collegio* of Milan at the time.[34] As we have seen in other parts of his writings, the author is very keen on listing people; in this case and in this specific place (the autobiography) he probably aims not only at making public the vast number of portraits he accomplished, but also at stressing the social privilege of knowing these people.[35]

These connections are important to point out especially if we consider the interrelations between the study of anatomy in the "official places" (such as the university of Pavia where the Carcano brothers studied) and the anatomical practices useful for the artist's work (such as the dissections made in the *ospedale militare* in Milan, where Giovanni Battista Carcano was employed until he became a professor). These places were accessible to external visitors –aside from the students of anatomy- under previous agreements, and we can legitimately speculate that Lomazzo, who was in contact with both brothers (as well as with other *dottori* of the city) thanks to his reputation as portraitist, could have had access to places like the *ospedale maggiore*, in which anatomical procedures were being performed.

Without doubt, studies of the human figure were copied in Lomazzo's *bottega*, as it was difficult to learn how to draw a body without a direct observation

---

1584). For the procedure Carcano was paid 25 *lire*. Quoted in Luigi Belloni, *Storia della Medicina a Milano* (Milan: Treccani, 1958–1962), p. 675. See also Nancy G. Siraisi, *Medicine and the Italian Universities* (Leiden: Brill, 2001), p. 363 ff.

31 Morigia, *La Nobiltà di Milano*, p. 155.

32 Picinelli, *Ateneo dei Letterati Milanesi*, p. 274.

33 Lomazzo, *Rime*, p. 541: "Col Carcan, Pietro Martire e il fratello/Ambi gran notomisti e gran chirurghi."

34 Lomazzo, *Rime*, pp. 541–542.

35 On the various functions of listing, see James Delbourgo, "Listing People," *Isis*, 2012, *103.4*:735–742.

of images (or models). While recipes for colors or notions on perspective (as we have seen in the previous chapters) were somehow easier to categorize theoretically, the study and the skill to reproduce the various parts of the figure could not be effectively put into words (even Lomazzo struggles to categorize them) and this subject required, arguably more than others, direct examination coupled by frequent copying of illustrations. That is one of the reasons why this learning process fostered collaborations with anatomists and doctors in order for artists to get the necessary bodily parts of which they could make drawings to use for realistic representations in their pictorial works.

The Anatomical Terminology in the Treatise and in the Codex Huygens: A Comparison

As we have briefly hinted at above, the terminology that Lomazzo uses for the different parts of the body has a strong colloquial flavor, as he lists some dialect terms and adds synonyms in *volgare* afterwards. Lomazzo adds three or more terms for each part of the body, such as for instance: "Almost under the chin is the knot called *groppo, gozzo, gutture* or *noce, collottola* and *nodo*."[36] The list of names exceeds the practical reasons for it, and can be attributed to the author's tendency to accumulate names and synonyms in order to be complete in his transmission of anatomical knowledge. The terminology used also exceeds Vesalius' description, and makes us speculate that the author might have also borrowed from a different source –perhaps one of the books of secrets with medical content circulating at the time. In fact, we will here compare Lomazzo's terms with the terms presented in the Codex Huygens. Both sources place the description of the human parts at the very beginning, that is in the part dedicated to proportion, with only one small difference: Lomazzo places the chapter just before explaining "the connections of the man of ten faces", while the author of the codex places them in fol. 3 (fig. 6.3), after explaining the numerical relationships between the different parts of the body (fol. 2, fig. 6.2).

Before starting the comparison, few more words on the Codex are needed.

The Codex Huygens (ca 1570), which was on show at and digitized by the Pierpont Morgan Library in 2013, is a collection of practical instructions probably used by artists in Lombardy in the second half of the sixteenth century. It is an intriguing document that collects previous sources in close connection with Leonardo's lost papers. The Codex, composed by fourteen rules or *regole*,

36 Ciardi, *Scritti*, II, p. 42: "Quasi sotto il mento il nodo, detto groppo, gozzo, gutture," or "noce, collottola e nodo." (Lomazzo, *Trattato*, p. 38).

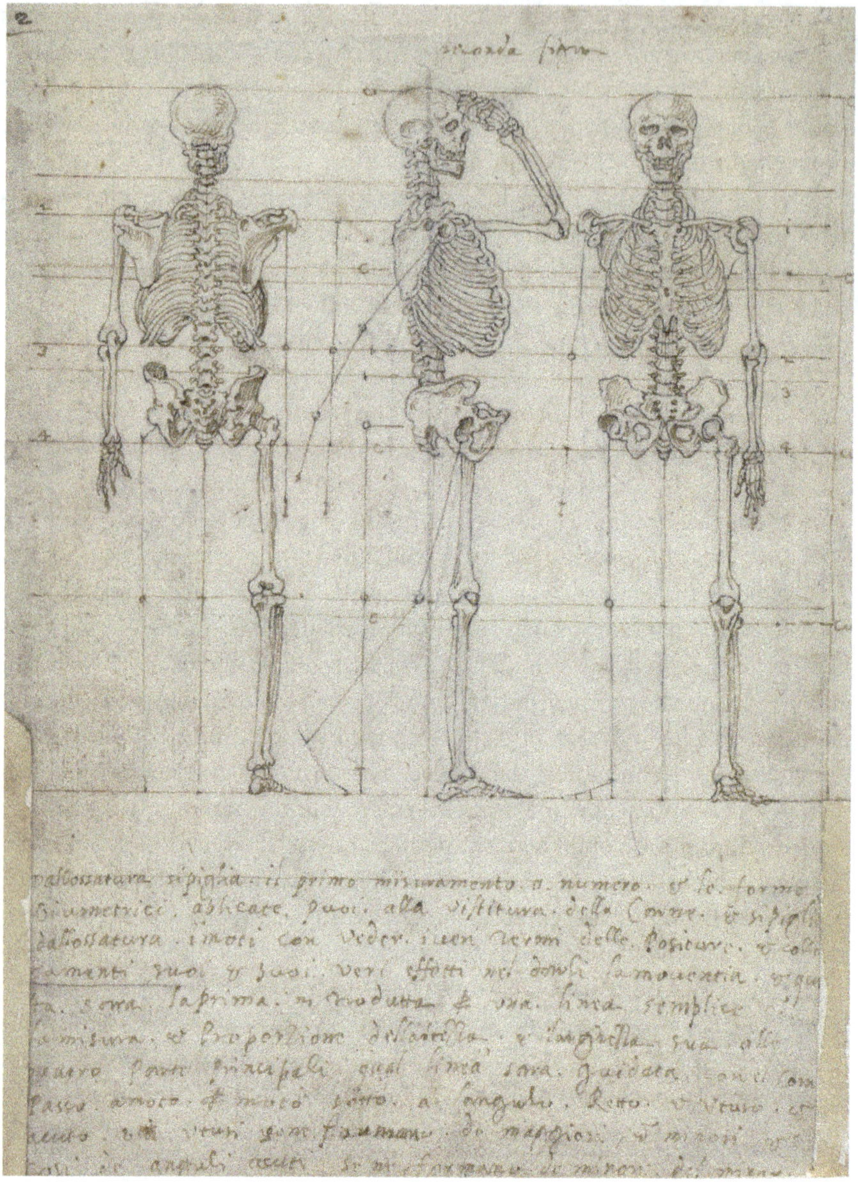

FIGURE 6.2 *Carlo Urbino,* Form and structure of the human body, *Codex Huygens, fol. 2, c. 1580, New York, Pierpont Morgan Library*

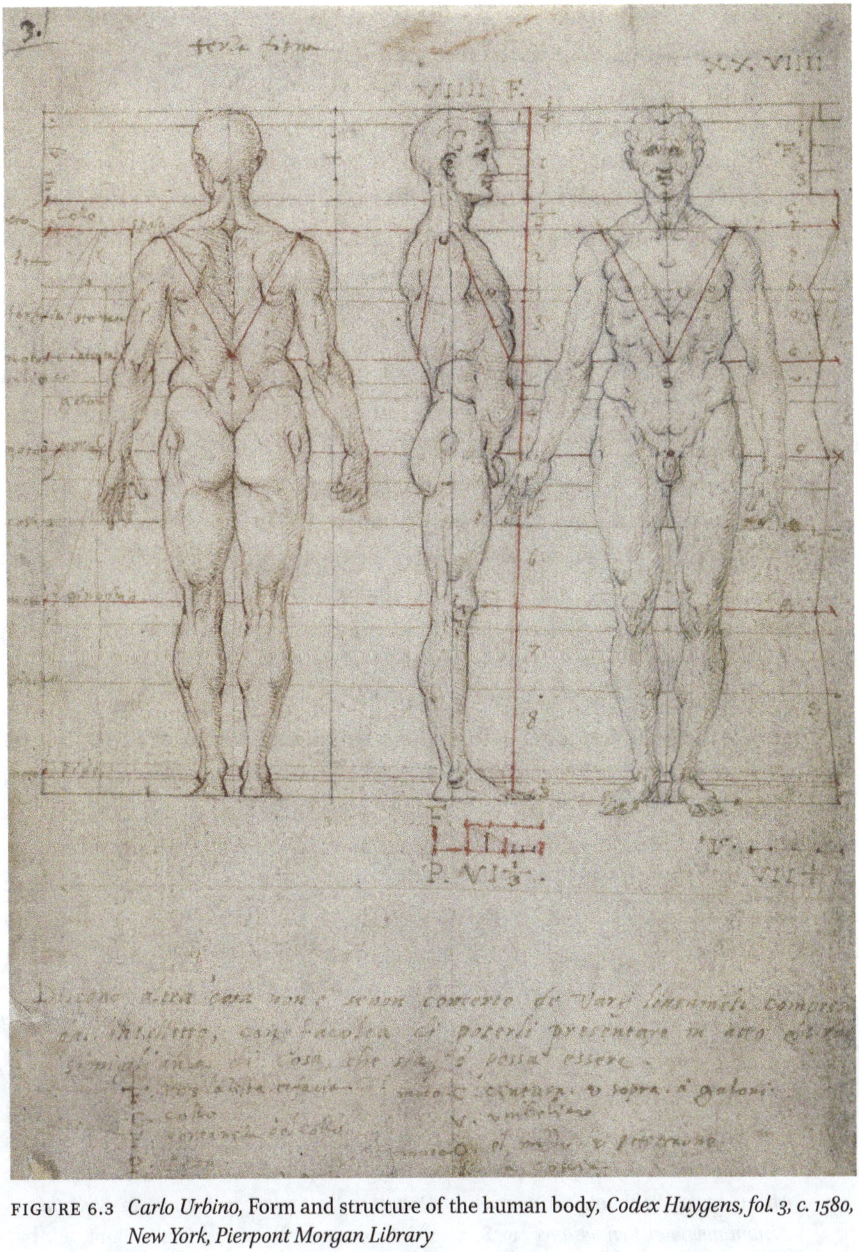

FIGURE 6.3 *Carlo Urbino,* Form and structure of the human body, *Codex Huygens, fol. 3, c. 1580, New York, Pierpont Morgan Library*

was mistakenly attributed to Leonardo, and subsequently to other artists such as Bernardino Campi, Aurelio Luini, Ambrogio Figino, and Lomazzo himself. The latest generally accepted attribution made by Carlo Pedretti is to the North Italian painter Carlo Urbino (ca. 1510/20–after 1585).[37]

The artist supposedly wrote the manuscript in order to make a treatise on painting and he was without doubt familiar with Leonardo's work, since many of the drawings seem to be copies of his now lost originals. This fundamental source is one of the many attempts to categorize pictorial knowledge on perspective, proportion and anatomy. Although it could be argued that this collection of instructions was not actually thought of as publishable in treatise form, the Codex, as we have already hinted at, has different points in common with Lomazzo's *Trattato*, both in its contents and in the organization of the subjects treated. It is therefore impossible not to mention it in the context of the study of the human body that was done at the time. The Codex has already been mentioned regarding the part on perspective, as it dedicates a long discussion on the *viste mentite,* which have been analysed before.[38] As for its structure, given the overall brevity of the written parts and the detailed schemes on the division of the body and on perspective and foreshortening constructions, we can speculatively assume that this writing was actually meant to be of use for artists in their pictorial work.

The manuscript is divided into five books, which include: Form and Structure of the Human Body, Movements of the Human Body, Transformations, Proportions of the Human Body –Proportions of the Horse and Perspective. We find a discourse on the description of the body mostly in books I and IV, which are always put into connection with (and preparatory to) the discourse on proportion, similar to Lomazzo's work. The similarities with Lomazzo's *Treatise* are more analogous in their categorization of the different parts of paintings and in their formal organization, rather than in content or purpose.[39]

---

37 See Bora, "La Prospettiva della Figura Umana" (op. cit.). The Codex was first transcribed by Panofsky, *The Codex Huygens*, who attributed it to Aurelio Luini. Other attributions were done by Irma A. Richter, "The Codex Huygens," *The Art Bulletin*, 1941, *XXIII*:335–338; Arthur Ewart Popham, "On a Book of Drawings by Ambrogio Figino," in *Bibliothèque d'Humanisme et Renaissance*, 1958, XX:266–276, especially p 274; Giulio Bora (ed. by), *Disegni di Manieristi Lombardi* (Vicenza: Neri Pozza, 1971), pp. 43–44, in which he attributed it to Lomazzo; and finally Carlo Pedretti, *The Literary Works of Leonardo da Vinci –A Commentary to Jean Paul Richeter's Edition* (Berkeley and Los Angeles: University of California Press, 1977), I, pp. 70–75.

38 See chapter 5.

39 As already noted by other scholars, as for instance by Marconi, "La Proiezione della Figura Umana", p. 183.

The terminology in the codex and in Lomazzo regarding anatomy is summarized in the following lists:

Codex Huygens (fols. 2 and 3)
Neck: *collo, fontanella del collo*
Chest: *petto*
Waist: *cintura*
Pelvis: *varchi*
Hip bones: *gallone*
Navel: *umbelico*
Crotch: *el mezzo, petignone*
Shin: *stinco*
Knee: *ginochio*
Ankles: *cavichie*
Root of the foot: *tacatura del piede*
Joints of the toe: *piegature*
Point of the shoulder: *node della spalla*
Elbow: *gomito*
Wrist: *chiave della mano*[40]

Lomazzo (*Trattato*, book I, ch. V)
Head: *testa, capo, zucca, sincipite, vertice, gnucca*
Parting of the hair and hair terminology:
scriminale, chime, crini, ciufetto, ciocca, crespi, abbellati, cuticagna
Ears: *orecchia, grasello, mirenga*
Eyes: *occhio, pupilla, acume, cornice degli occhi, cassa degli occhi*
Nose: *naso, nari, narici, forame, papilla*
Parts of the mouth: *bocca, canaletto, palato, lingua, canale, pallone, gingiva*
Teeth: *denti, tomis, canini, mascellari, radici*
Chin: *mento, barbozzo, groppo, gozzo, gutture*
Chest: *petto*. For women: *mamma, cizzo, ubere, pomo, poppa, zinna*
Nipple: *bollino, pupilla, capitello*
Stomach: *stomaco, casso e torace*
Underarm: *ascella, lesena, ditella*
Waist: *costato, cintura*
Crotch (female): *pancia, ventre, ventraia, alvo, utero, pettinecchio e pettignone, natura, vulva, conno, fica, fessa*
Crotch (male): *naturale, verga, membro, coda, piuolo, pisello, pinca, caviglia, priapo*[41]

40 The Codex's content is also summarized in Panofsky, *Codex*, pp. 20–21.

41 Ciardi, *Scritti*, II, pp. 41–44.

For the hand, Lomazzo adds to the terminology of each finger his corresponding planet, that is:

Venus: thumb
Jupiter: index
Saturn: medium finger
Sun: ring finger
Mercury: little finger
Moon and Mars (mountains of): superior and inferior palms.[42]

As we can see, Lomazzo uses different terms to define each part, many with a strong dialectic flavor (*casso, barbozzo, cuticagna*), which could be taken either from the reality of the Milanese workshop or, possibly, from the burlesque literary tradition Lomazzo was so well acquainted with. On the other hand, we see that the beginning of the Codex Huygens underlines more the importance of the study of the human body as a whole, since the first image in Book I presents a detailed representation of the human skeleton with three skeletons in different standing poses, which are then reproduced in the following folio, where three studies of human figures in the same positions are depicted "dressed with the flesh."[43] The terminology used in the Codex is synthetic and not vastly elaborated, as are the illustrations that accompany it, since in this case the author's main interest concerning the human body is to understand the proportions between the various parts. In the codex, schematic drawings of the hands, fingers, arms and feet are presented in order to transmit the right dimensions and relationships that link all the parts with the aim of constructing harmonious figures. The explanations are for the most part strictly mathematical, and the purpose evidently shifts from copying detailed anatomical parts (as Leonardo's drawings were meant for) to recreating the overall harmony of the body in all its parts. The process of categorizing anatomical parts presented in the Codex Huygens is likely a product of that visual culture promoted by artists who related their interest in the human body with workshop's practices. However, when we compare the content of this with Lomazzo's discourse presented above, one cannot deny that the discourse of the latter differs from the schematic categorization of the former, as Lomazzo elaborates

42 Ciardi, *Scritti*, II, p. 43.

43 Codex Huygens, folio 2: "Dall'ossatura si piglia il primo misuramento e numero e le forme metriche. Applicate poi alla vistitura della carne e si pigli dall'ossatura i moti."

with many synonyms on the different parts of the body. This is also evident in the parts dedicated to anatomy in *Della Pratica della Pittura* (although from the title one may suppose the contrary), since two long chapters on the shape of the human body and of its bones are mostly descriptive and have the tendency to accumulate names of painters skilled in depicting properly the human figure. This gave readers *exempla* of good artists from which to copy and to admire, constituting therefore a canon of skillfullness, if not beauty. We have as usual long lists of names: "Of the painters the most excellent were only Raphael, Perino del Vaga, Rosso Fiorentino, Marco da Siena, Salviati, Pellegrino Pellegrini, Giovani Fiamengo who drew the anatomy of Vesalius, and Aurelio Luino."[44]

Lomazzo interprets the works of previous and contemporary painters, listing them and praising especially the works of the painters whom he knew personally (such as, predictably, Aurelio Luini). Moreover, when describing the bones, the author seems mostly interested in giving the number of all of them: "The head is composed by eight bones, the first occupies the forehead, the second and the third make the lobes, the fourth and fifth occupies the temples."[45] These long lists constitute a body of knowledge that contributes to the overall scope of the book, that is to educate readers in the art of painting by giving them both practical instruction and judgements on the art of the time. Lomazzo seems aware of the difficulty of talking about anatomy without direct visualization of the parts of the body, and that is the reason why he probably does not "teach" the reader how to depict the parts that he mentions. What we find is a scattered discourse on the depiction of the body, while the relationships between the different parts are extensively explained in the first book on proportion. At the end of chapter XXIV of *Della Pratica,* he admits that: "This is what I wanted to sum up here as briefly as possible (*the composition of the bones*), because I have enquired on the muscles and on their function elsewhere. But to be completely honest, to understand them perfectly, anyway

44 Ciardi, *Scritti*, II, p. 534: "De i pittori soli sono stati eccellenti Rafaello d'Urbino, Perino del Vaga, Rosso Fiorentino, Marco da Siena, il Salviati, Pellegrino Pellegrini, Giovani Fiamengo che disegnò l'anatomia al Vesaglia, e Aurelio Luino." (Lomazzo, *Trattato*, p. 615).

45 Ciardi, *Scritti*, II, p. 535: "Il cranio che si compone di otto ossa, il primo occupa la fronte e da lui piglia il nome, il secondo e terzo fanno la coronella, il quarto e quinto occupano le tempie, (…) il sesto piglia la collottola, il settimo si incassa." (Lomazzo, *Trattato*, p. 616).

it is necessary to see them from life, such as good painters and sculptors have done thousands of times."[46]

In order to see which kind of visual material Lomazzo meant, I will now focus on a series of anatomical drawing attributed to a colleague of the author, the aforementioned Annibale Fontana.

Annibale Fontana's Studies of Bones: The Circulation of Anatomical Drawings from Master to Pupil and in the Blenio Academy

Annibale Fontana was a renowned Milanese sculptor and engraver, a friend of Lomazzo who also took part in the *Accademia de la Val di Blenio*. Fontana was defined by Lomazzo as: "Marvelous sculptor not only in Blenio, but in all the other countries (...) magisterially skilled in carving crystals and precious stones, as well as in carving marble and statues."[47] The author probably knew him from an early age, because he mentioned his name in the *Libro de' Sogni* stating that: "Although he is so young, he is performing miracles in the art of jewelry, with great effort and study", and because Fontana carved a medal taken from the portrait of a young Lomazzo now in Vienna.[48]

Nine anatomical drawings from to the Biblioteca Ambrosiana are attributed to this artist (figs 6.6–6.11), and they bear striking resemblances to Leonardo's anatomical illustrations collected in the Windsor folios of the Royal Library. Leonardo's folios are dated around 1510 –following an annotation on the recto of folio 19016, in which the artist states that he expects "to finish all the anatomies in winter"- and they constitute the most important testimony to the artist's interest in the human body, not only in a descriptive but also in a mechanical sense, as the artist conceived of the body as an active mechanical

46 Ciardi, *Scritti*, II, p. 536: "Questo è ch'io, quanto più brevemente ho potuto, ho voluto raccorre qui della composizione dell'ossa; perchè de muscoli e degl'offizi loro ho ragionato a bastanza altrove. Ma per dirne liberamente quel ch'io sento, per intendergli perfettamente, ad ogni modo è necessario vedergli dal vero siccome hanno fatto mille volte i buoni pittori e scultori." (Lomazzo, *Trattato*, p. 620).

47 Isella, *Rabisch*, p. 65: "E fu facc dar compà Rivebed Fontana (nominad scià inanz ch'o l'intrass in sta Vall Annibal Fontana) scultó maraviglios no solament in Bregn, ma in tucc gl'olter paglis, si com de scavà sciò in dor crestall e in di pregl' preciglios sciò che voreva."

48 See Ciardi, *Scritti*, I, p. 166: "Annibale da Milano che, per esser sí giovanetto, fa nell'arte dell'oreficeria miracoli con sommo studio e fatica."

apparatus.[49] Folio 19011 recto (fig. 6.4) shows a so-called "exploded view" of the leg and the foot, in which the artist demonstrates how the bones of the lower legs (tibia and fibula), are articulated to the bone of the foot (talus). Then, on another folio we have studies of the muscle of the arm and of the shoulder, and at the bottom right the profile of the foot again (folio 19013 verso, fig. 6.5).

In the annotation, Leonardo writes:

> Make the bones of the foot somewhat separated from one another, in such a way that one can distinguish distinctly one from the other; and this is the way to give true knowledge of the number of the bones of the foot and of their shape.[50]

On another sheet we have legs set apart to see the muscles clearly, as explained in the note:

> Make a demonstration with lean and thin muscles so that the space that appears between the one and other makes a window to demonstrate what is behind them – like this drawing of a shoulder made here with charcoal.[51]

These and many other illustrations were notably collected by the pupil Francesco Melzi (1491–1568), who made them accessible to artists who wished to see them. They were later acquired by the Milanese sculptor Pompeo Leoni (1533–1608), who bought most of the folios possessed by Melzi's nephew, among which are the anatomical drawings now in Windsor.[52] It is perhaps worth mentioning here that the father of Pompeo, Leone Leoni (1509–1590), sculptor and collector, had knowledge of Lomazzo's work and literary achievements, as a letter by the unknown Valerio Angelini to Giovanni Paolo testifies

49 See Kenneth Clark and Carlo Pedretti, *Leonardo da Vinci's Anatomical Drawings at Windsor Castle* (London: Phaidon, 1969). For a discussion on Leonardo's conception of man as a mechanical being, see among others Paolo Galluzzi "Leonardo da Vinci: From the 'Elementi Macchinali' to the Man Machine," *History of Technology*, 1987, 4:235–265.

50 Keele and Pedretti, p. 546, in Cazort, Kornell, Roberts *The Ingenious Machine of Nature.*

51 Keele and Pedretti, p. 560, in Cazort, Kornell, Roberts *The Ingenious Machine of Nature.*

52 For the fortune and later dispersion of the drawings, see among other studies the recent book by Carmen Bambach, *Un'Eredità Difficile: i Disegni ed i Manoscritti di Leonardo tra Mito e Documento*, (Florence: Giunti, 2009) and Claire Joan Farago, *The Fabrication of Leonardo da Vinci's* Trattato della Pittura, *1651* (forthcoming).

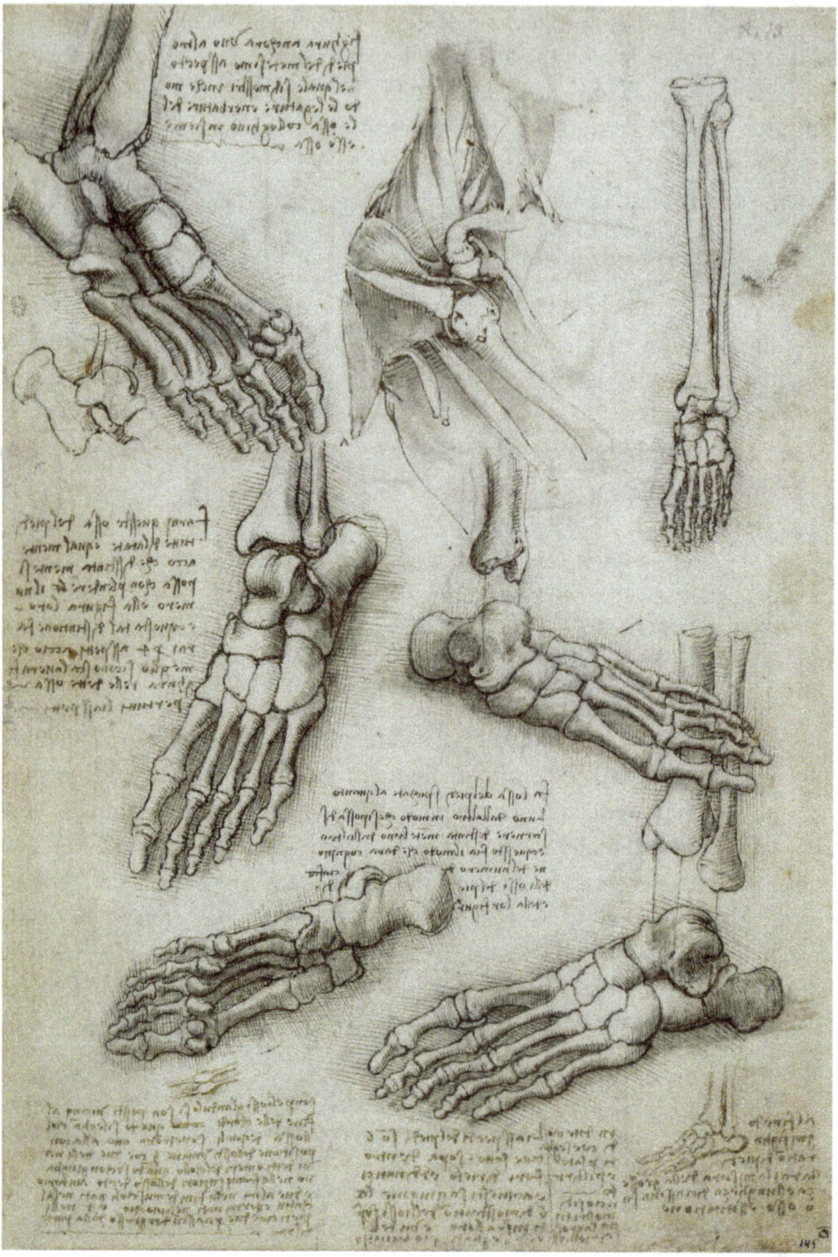

FIGURE 6.4 *Leonardo da Vinci,* Studies of the foot and shoulder, *folio 19011r, c. 1510, Windsor, Royal Library*

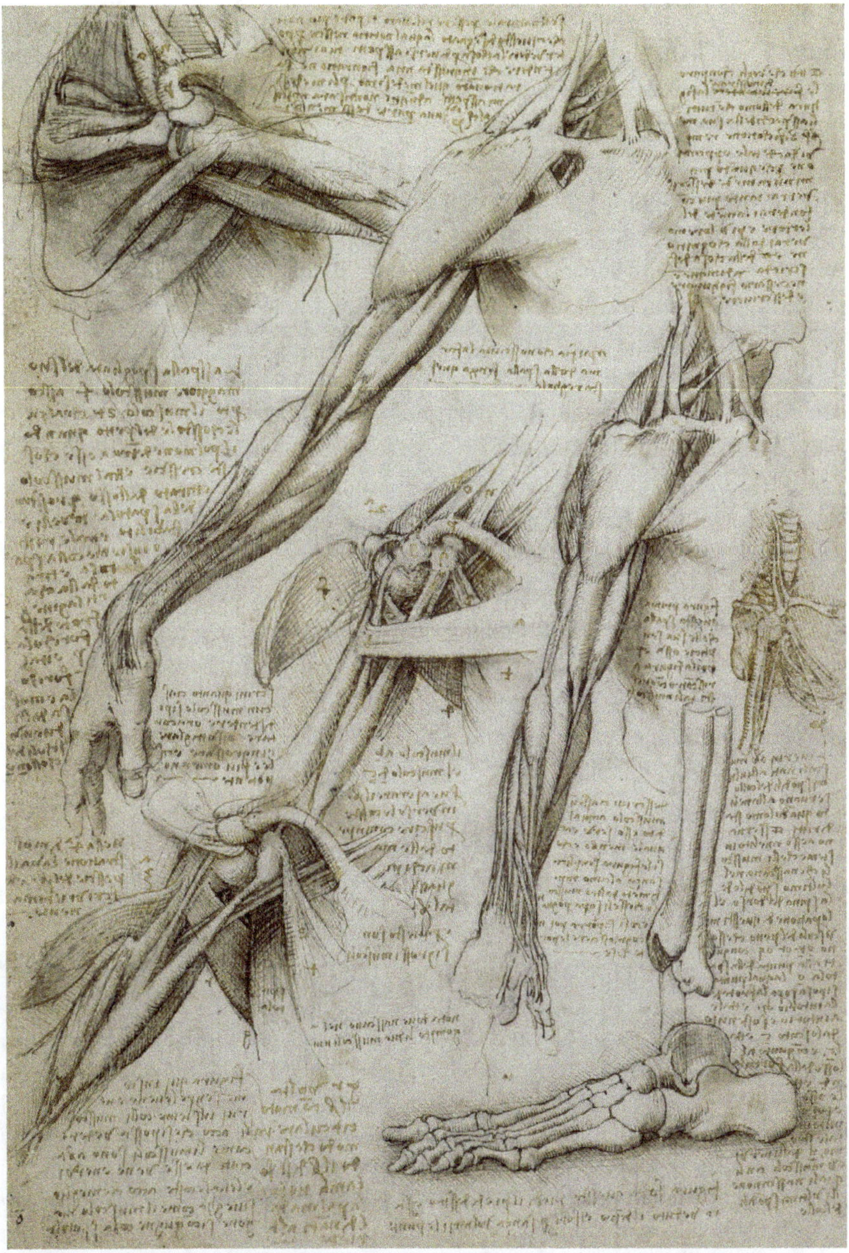

FIGURE 6.5 *Leonardo da Vinci,* Studies of the arm, shoulder and foot, *folio 19013v, c. 1510, Windsor, Royal Library*

to. The letter is dated 29th October 1586 (two years after the publication of the *Trattato*) and Angelini writes to the author that he is contacting him since "the honourable Leone Leoni celebrated your egregious name many times."[53] Leoni, having lived and worked in Milan for years, probably had the chance to meet Lomazzo personally, given the fact that they had an acquaintance in common: the literate Giuliano Goselini, who, as we have seen, was member of Blenio and praised Lomazzo's literary work in his *Rime di Diversi Celebri Poeti dell'Età Nostra*.[54] Leone and Goselini met in the context of the prestigious *Accademia dei Fenici* in Milan, probably around 1550, when the latter became secretary of the new governor Ferrante Gonzaga.

Whatever the pattern of these connections, it is certain that before Leoni's son bought the collection from Melzi and brought the folios to Spain (where he moved most likely at the end of the sixteenth century), these fundamental anatomical images were circulating in Milan among painters. Made available to consult by Melzi himself, they were undoubtedly known in the circle of the artists belonging to the *Accademia di Blenio*.

As we have seen in the chapter dedicated to it, Annibale Fontana was member of the academy under the name of *Compà Ribeved, scultó dra Val* (fellow Ribeved, sculptor of the valley). The anatomical illustrations in the Ambrosiana which are attributed to him consist of nine drawings: two made with black chalk on blue paper, beautifully sketched with faint touches of white chalk, and seven made with black chalk on white paper, some of which present some stains on the upper right. The latter drawings are qualitatively inferior to the first two drawings, which could lead one to think that only the ones on blue paper are actually by Fontana, and the others might have been copies after.[55]

---

53 Letter of Angelini published as appendix in Lomazzo's *Rime*. See the General Conclusions.

54 Goselini, *Rime*, p. 239: "Chi può cieco chiamar un ch'a l'oblio/Tolsi d'eterna notte? Un, ch'Argo al Cielo/Vola? Di cui non hebbe il mortal velo/O madre antica, figlio unqua si pio?/Cieca, e muta giaceasi, ed aspro, e rio/Sentia già del suo fine il freddo gielo,/Del pinger l'arte; e'l costui caldo zelo/Gli occhi, e le labra a lei languente aprio:/C'hor parla, e mira, e per lui vive, e'nsegna:/Et ei, fatto per lei maestro egregio,/La sua pietà con doppia Gloria illustra./Già del pennello, hor de la penna ha'l pregio:/Né de la Cetra sua quel Dio lo sdegna,/Che pingendo, e cantando il mondo lustra."

55 It is quite difficult to establish without further technical investigation, although I believe that the style and the organization of the sketches both on blue and white paper are similar enough to suppose that they were all made by the same hand. I thank Professor Giulio Bora for discussing the attribution of the drawings made by Anna Patrizia Valerio. See Valerio, "Annibale Fontana e il Paliotto dell'Altare della Vergine dei Miracoli in Santa Maria Presso San Celso," *Paragone*, May 1973, *24.279*:32–53, p. 48, n26.

The representation of the foot's bones (fig. 6.6) is quite similar to the one in the Windsor Folios presented above (fig. 6.4 and 6.5), as is the depiction of the arm's muscles on folio 25 (fig. 6.7). Moreover, the composition of the fibula and tibia bones on the upper right of folio 19011 (fig. 6.4, upper right) is almost identical to the composition of the bones in folio 22 (fig. 6.8), where those bones are reproduced three times (left and right legs).

The similarities in style and organization of these pages with Leonardo's folios are too evident to be ignored.

As Pedretti stated, Leonardo's anatomical drawings belong to a later phase of his studies, and he might have started them around 1510–1512. Lomazzo also mentions the name of Melzi when he talks about the form of the human body (*Trattato*, book VII, ch. XXIV): "After him (*Michelangelo*) Leonardo da Vinci was excellent in anatomy, and we find different drawings in the house of Francesco Melzi, a Milanese gentleman, his pupil, and the anatomy of the horse made by him."[56]

It is very likely that these drawings were circulating among artists who were acquainted with one another thanks to their participation in the academy, and no longer solely the province of master-pupil relationships.

We can infer that the circulation of these drawings gradually became wider, as they started to be seen and studied not only in the private sphere of the workshop, but also in the (still private to some extent, but certainly more public) sphere of the Blenio academy.

Of course, the master-pupil relation was still active and fruitful: if we look at the gesture of the arm and hand with the index finger lifted up in four drawings (fig. 6.9, 6.10, 6.11 and 6.12), which is quite a common iconographic gesture in the depiction of saints, we see that these representations are directly linked to at least two contemporary anatomical drawings by Ambrogio Figino, the most renowned pupil of Lomazzo at the time. The anatomical illustration I present here (fig. 6.13), sold by Sotheby's nine years ago, shows the same composition and a similar style in the depiction of the bones and skull, making it possible to draw a connection with the other drawings presented here and attributed to Fontana.

56 Ciardi, *Scritti*, II, p. 533: "Doppo eccellenti di lui sono stati Leonardo da Vinci, del quale si ritrovano diversi disegni in più mani, e principalmente in casa di Francesco Melzi, gentiluomo Milanese, suo discepolo, oltre l'anatomia dei cavalli che egli ha fatto." (Lomazzo, *Trattato*, pp. 614–615). For a recent discussion on the anatomical studies of Leonardo, see Pedretti, "Il Corpus degli Studi Anatomici di Leonardo a Windsor fra Bologna e Los Angeles," in Giuseppe Olmi (ed. by), *Rappresentare il Corpo: Arte e Anatomia da Leonardo all'Illuminismo* (Bologna: Bononia University Press, 2004), pp. 19–30.

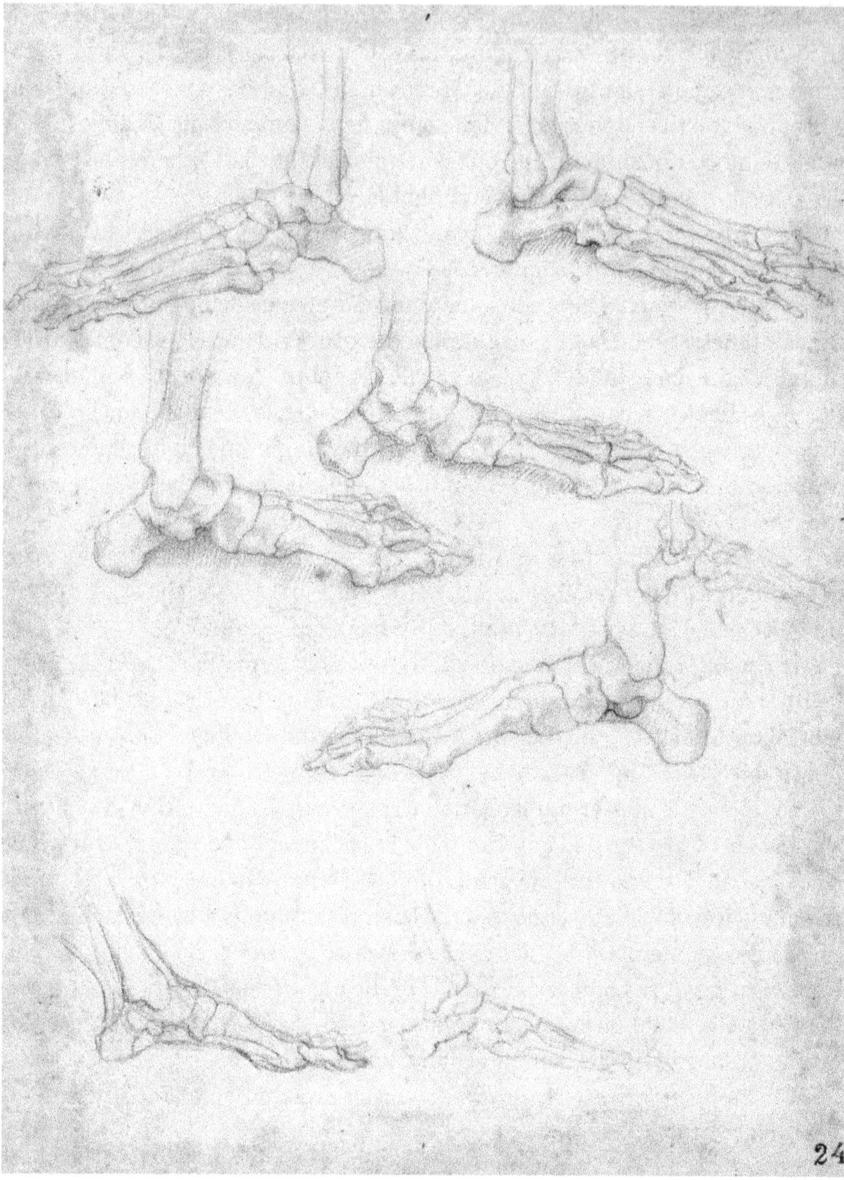

FIGURE 6.6 *Annibale Fontana (attributed to),* Eight skeletal studies of a human foot, *F 245 folio 24,* ND *cat. No 41, c. 1550, Milan, Biblioteca Ambrosiana*

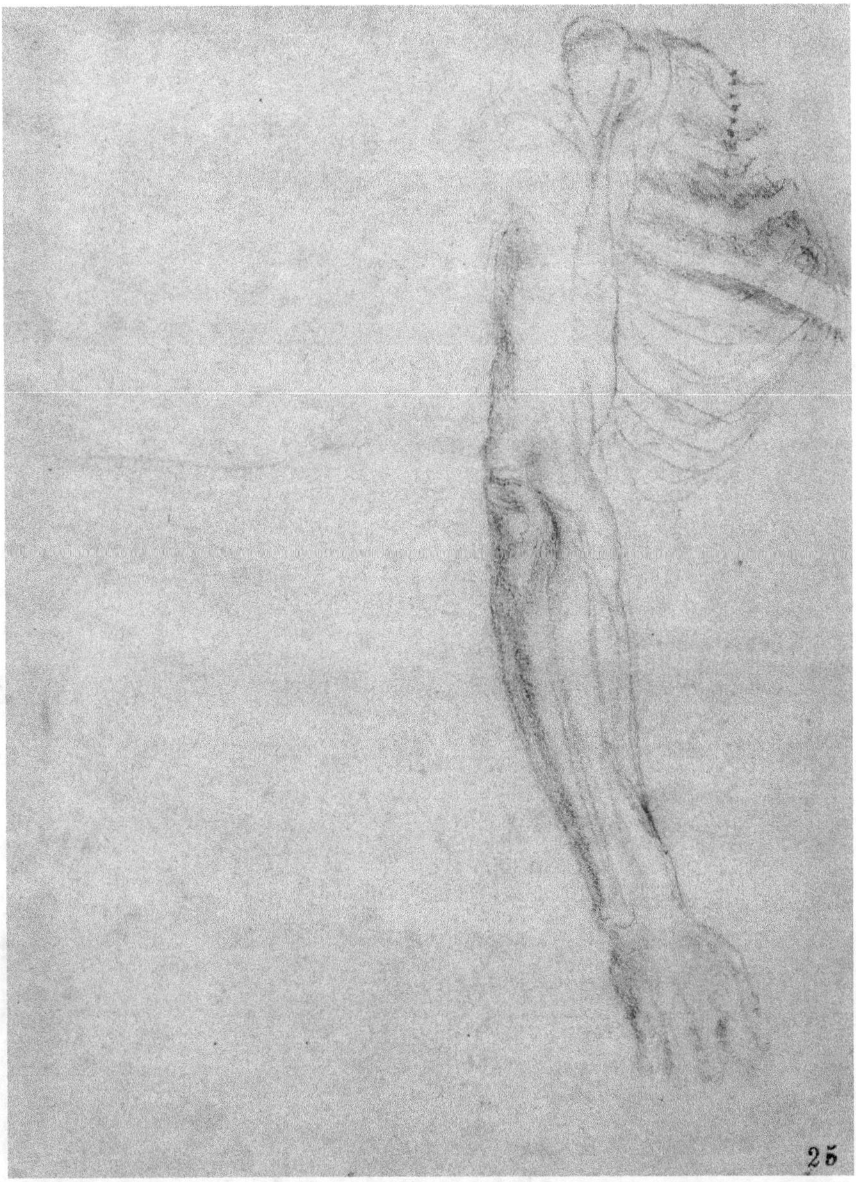

FIGURE 6.7 *Annibale Fontana (attributed to),* Right side, a human right arm and rib cage with muscles and bones, *F 245 folio 25,* ND *cat. No 42, c. 1550, Milan, Biblioteca Ambrosiana*

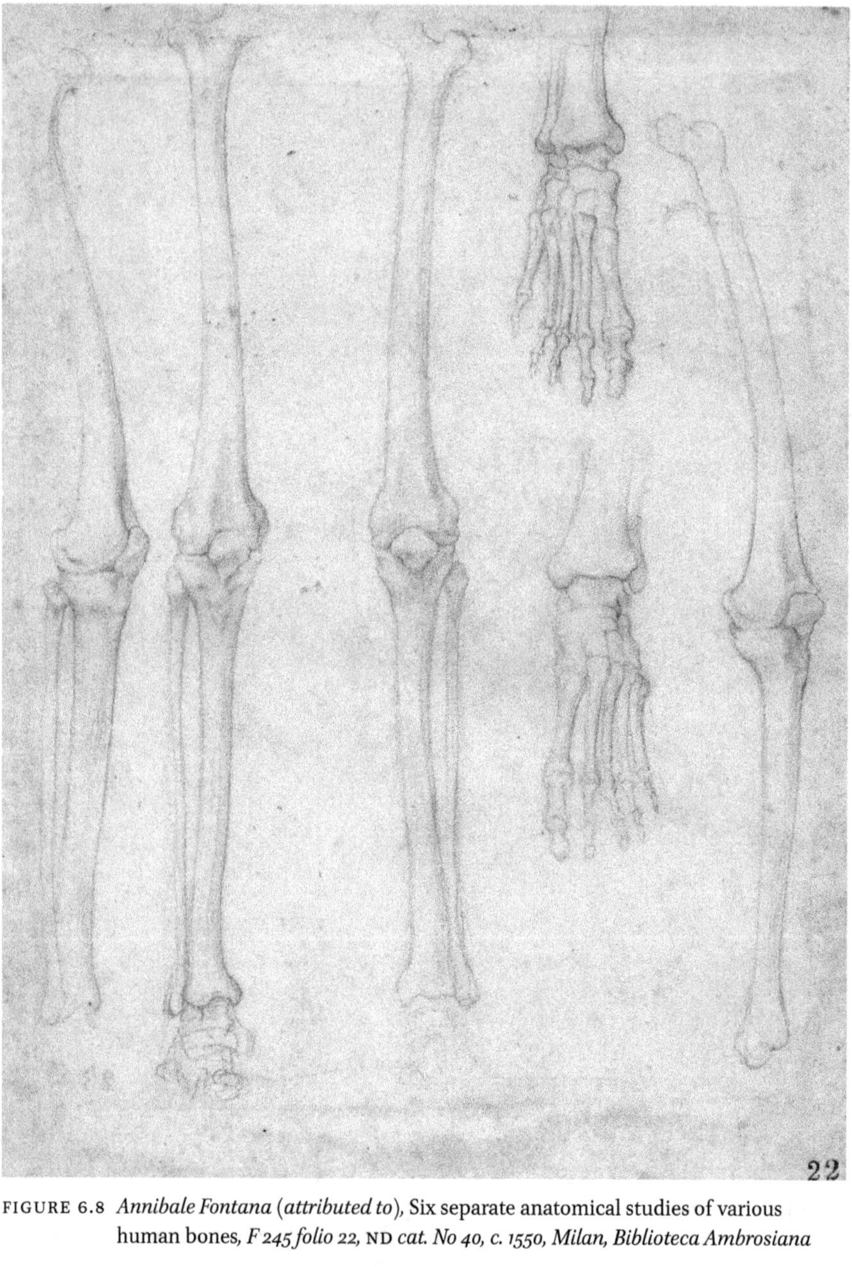

FIGURE 6.8 *Annibale Fontana* (*attributed to*), Six separate anatomical studies of various human bones, *F 245 folio 22,* ND *cat. No 40, c. 1550, Milan, Biblioteca Ambrosiana*

Figino himself was quite skillful in making studies of the human muscles, as many other drawings show, like the ones in the Royal Library of Windsor, the Victoria and Albert Museum, or the sketches surrounding the famous *Nude Demon Encircled by a Serpent* at the Metropolitan Museum of Art. On the left (both upper and lower) side of the latter drawing we have a series of studies of the legs muscles and few heads, similar in style to Leonardo's studies.

As Domenico Laurenza convincingly showed, Figino directly used some of Leonardo's lost sheets as models for his drawings of comparative anatomy, which came into the artist's possession thanks to his working relationship with Lomazzo (and with respect to one lost manuscript, also thanks to Giovanni Ambrogio Mazenta, the brother of Guido Mazenta, a Milanese man of letter and another admirer of Lomazzo's work).[57] The artist himself, declaring, as we have already stated, that he possessed "four thousand papers and more", probably owned some of the Leonardo sheets that were subsequently dispersed.[58]

The illustrations presented so far, which are all linked to Lomazzo's circle of artists, testify to the widespread practice of copying drawings in order to appropriate the knowledge of human anatomy (and of the anatomy of the horse, which I deliberately decided not to include here), with the aim of making a lifelike depiction of figures in paintings, sculptures or medals.

The study of anatomy theoretically could have been done directly, analyzing dissected bodies or models, or indirectly, by copying illustrations and drawings such as Leonardo's. We find a visual representation of the "direct learning" in the background of the drawing *The Allegory of Painting* (fig. 6.14), which is attributed to Lomazzo and belongs to the Albertina in Vienna. While the subject on the foreground constitutes a metaphor for the Art of Painting, the background shows three different scenes of artists' workshops as they could have been in reality.[59] This background is in itself of extreme interest, as it

57 Domenico Laurenza, "Figino and the Lost Drawings of Leonardo's Comparative Anatomy," *The Burlington Magazine*, March 2006, *148*:173–177. In this article we also find another illustration that is Figino's various studies of the human body, where we have the same gesture of the arm with the index finger lifted up repeated two times at the very centre of the drawing,.

58 I agree with Laurenza, "Figino and the Lost Drawings", p. 177, n23.

59 On the interpretation of this allegory see for instance James B. Lynch Jr, "Lomazzo's Allegory of Painting," *Gazette des Beaux Arts*, 1968, *110*:325–330, and Catherine King, "Late Sixteenth-Century Careers' Advice: a New Allegory of Artists' Training – Albertina, inv. No. 2763," *Wiener Jahrbuch für Kunstgeschichte*, *XLI.1*:77–96. The latest interpretation of this drawing was done by Prof. Ulrich Pfisterer in his book *Kunst-Geburten: Kreativität, Erotik, Körper* (Berlin: Verlag Klaus Wagenbach, 2014), pp. 65–68. I thank Prof. Wolf Dietrich-Löhr for the suggestion.

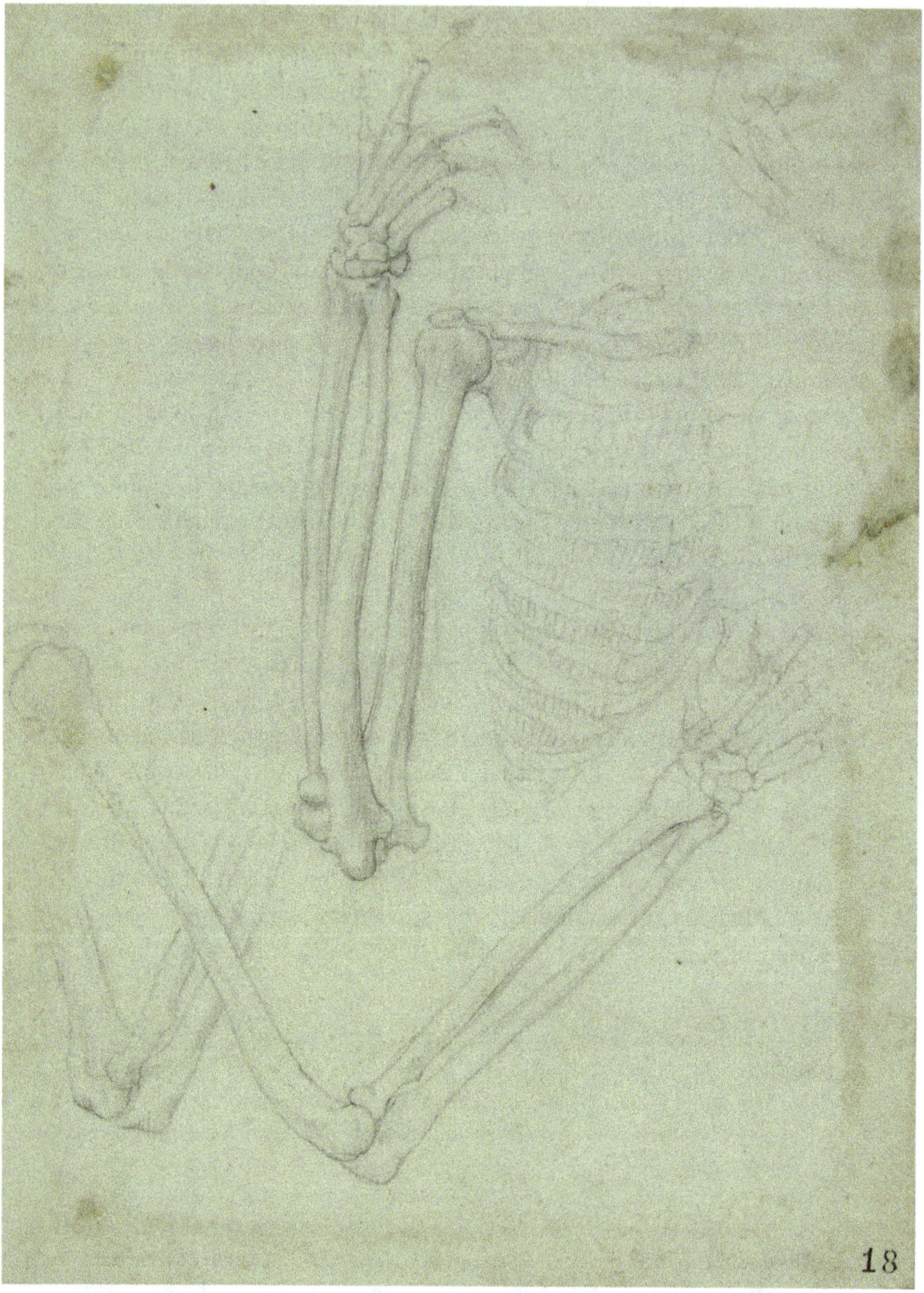

FIGURE 6.9 *Annibale Fontana* (*attributed to*), Various anatomical studies of human bones and muscles, *F 245 fol. 18,* ND *cat. No 36, c. 1550, Milan, Biblioteca Ambrosiana*

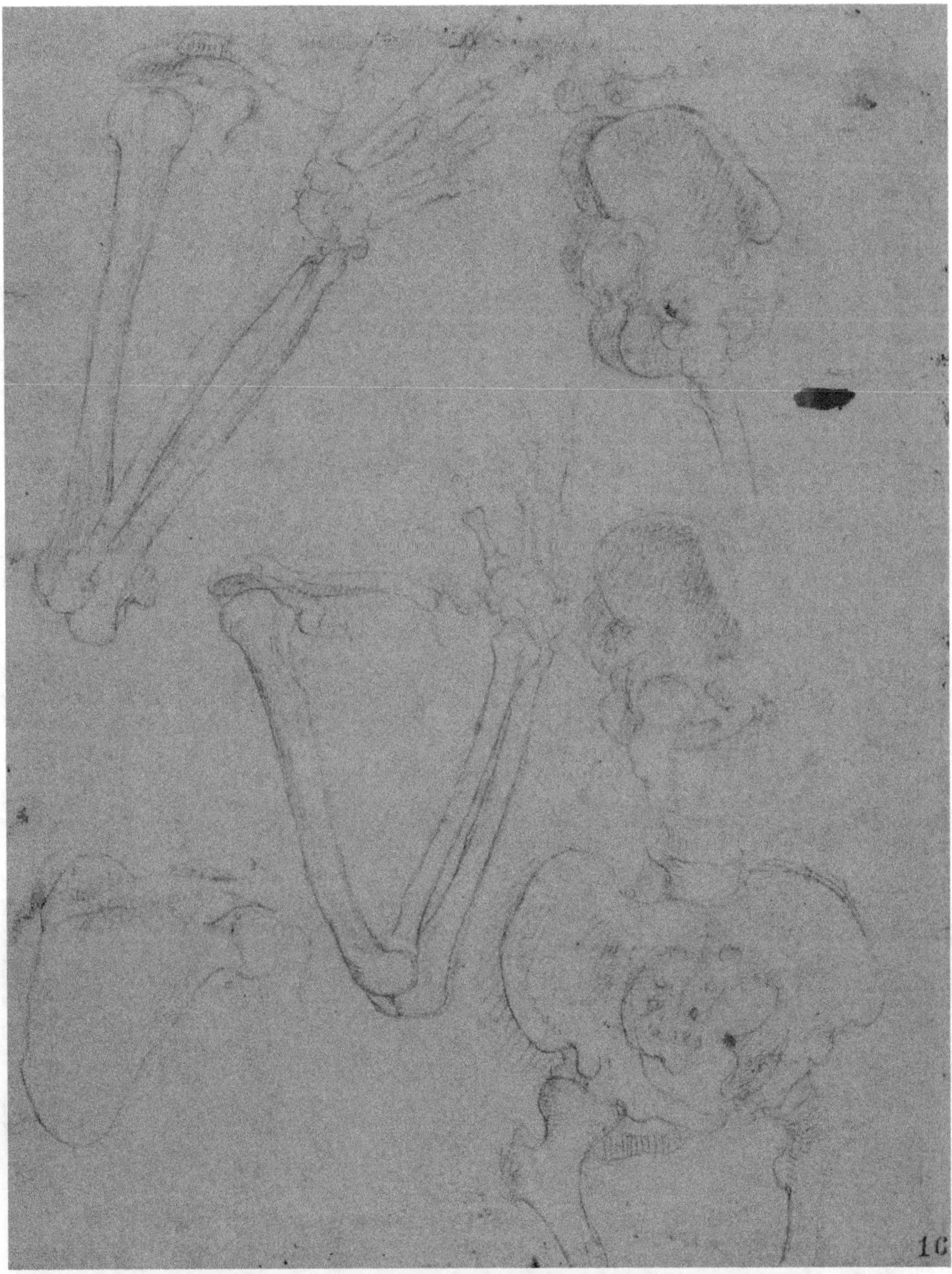

FIGURE 6.10 *Annibale Fontana (attributed to)*, Three separate anatomical studies of various human bones, *F 245 fol. 16,* ND *cat. No 34, c. 1550, Milan, Biblioteca Ambrosiana*

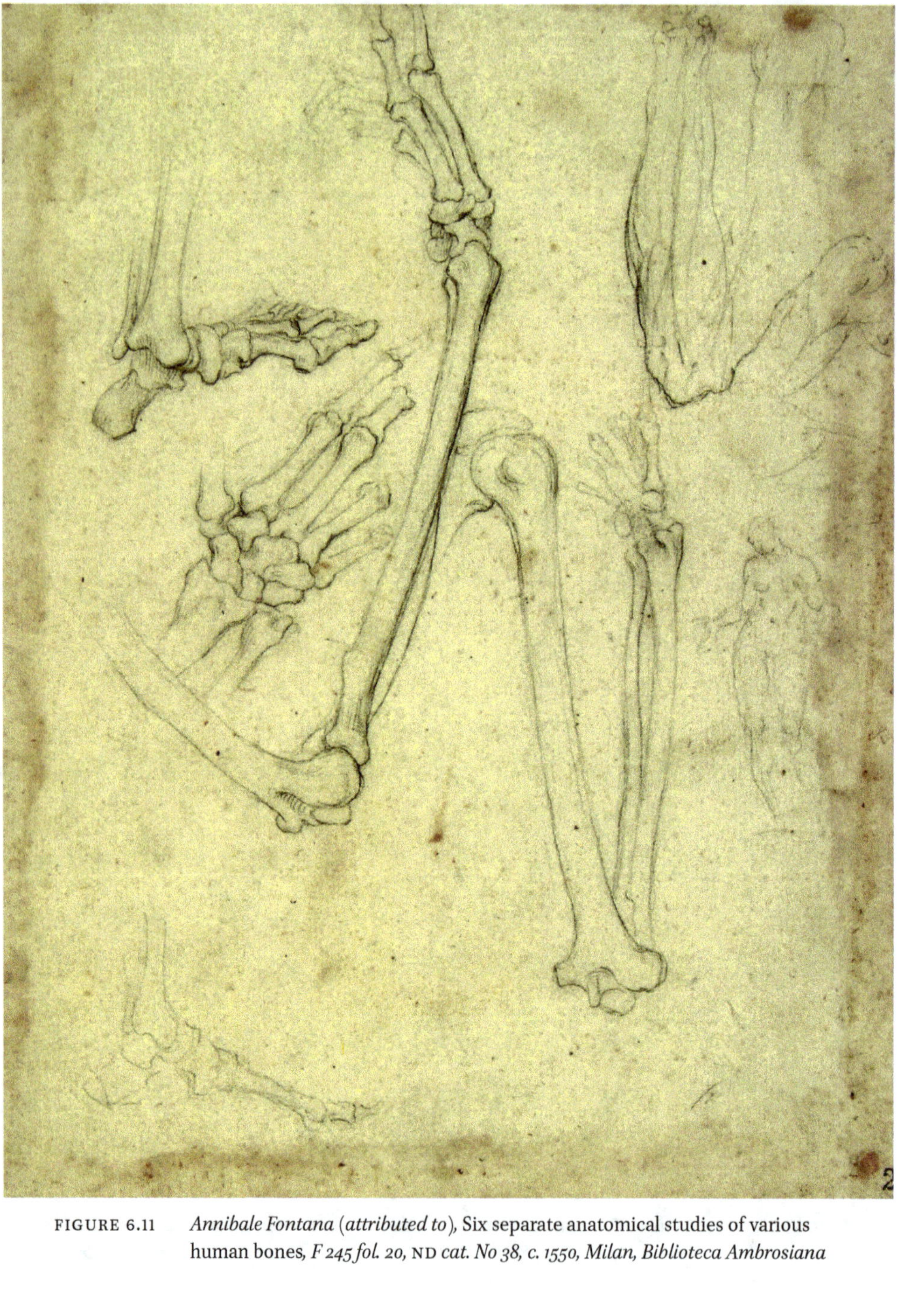

FIGURE 6.11 *Annibale Fontana (attributed to)*, Six separate anatomical studies of various human bones, *F 245 fol. 20,* ND *cat. No 38, c. 1550, Milan, Biblioteca Ambrosiana*

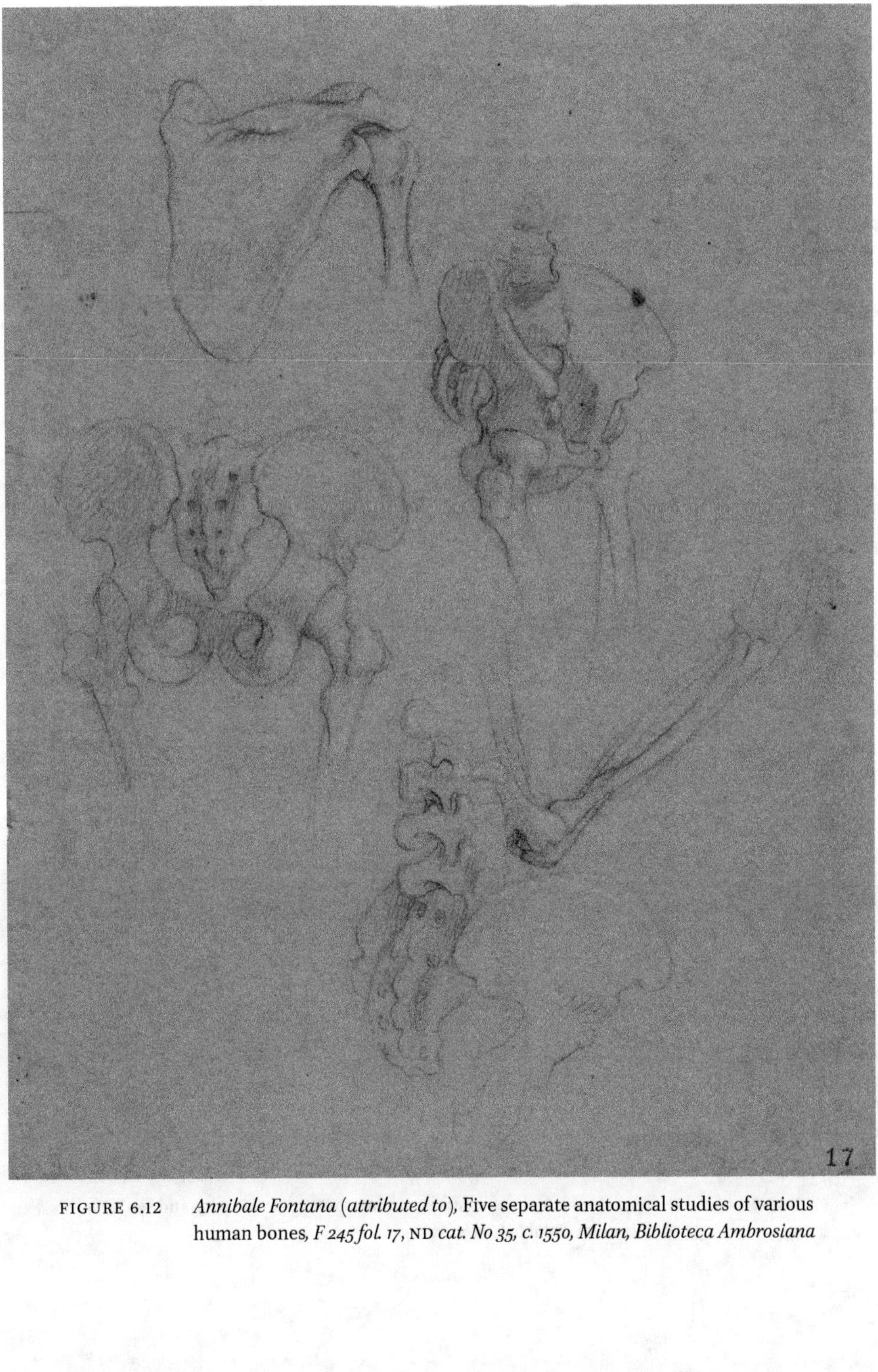

FIGURE 6.12 *Annibale Fontana* (*attributed to*), Five separate anatomical studies of various human bones, *F 245 fol. 17*, ND *cat. No 35, c. 1550, Milan, Biblioteca Ambrosiana*

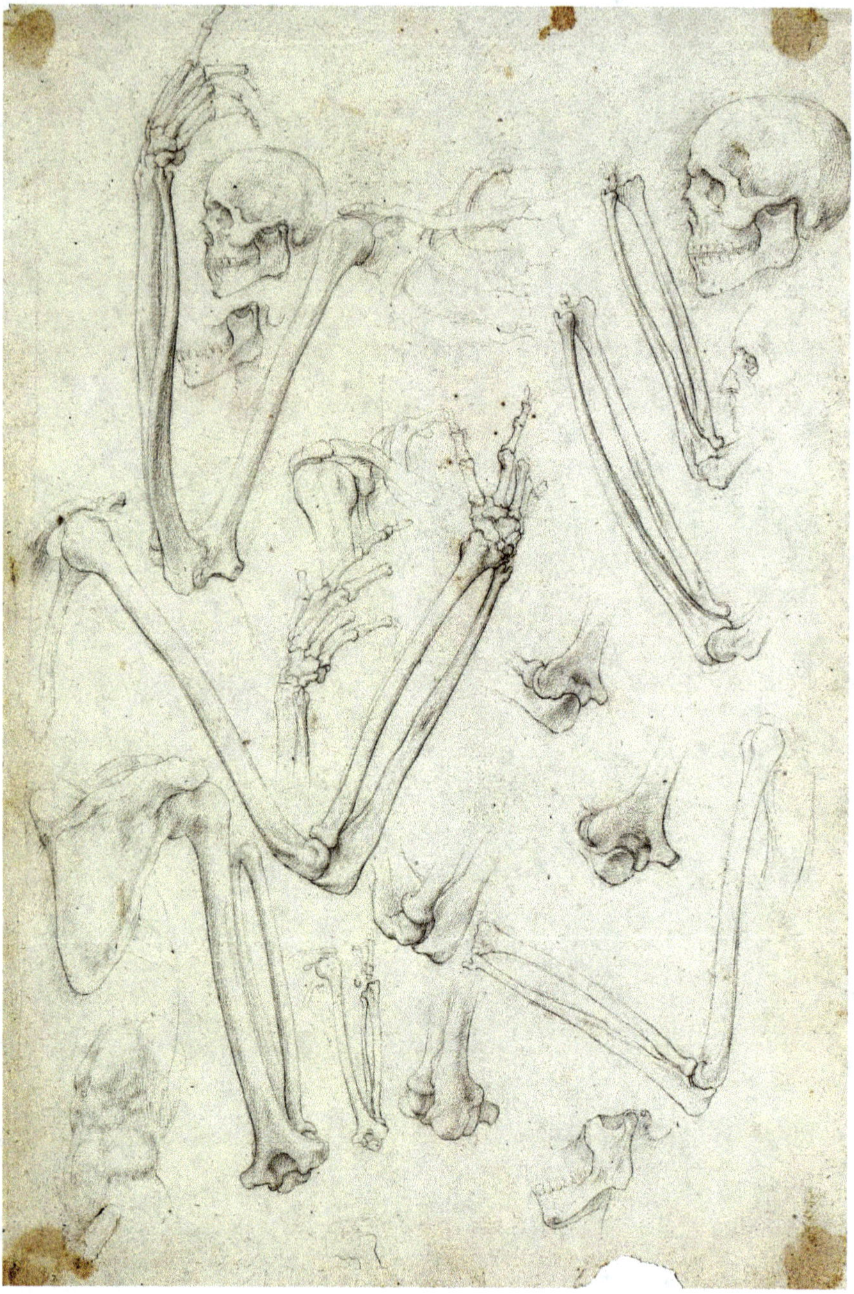

FIGURE 6.13 *Ambrogio Figino,* A Sheet of studies of human skeletons: arms and skulls, *c. 1568, London, Sotheby's, July 4, 2007*

gives us one idea of the daily routine in the *bottega*. On the right side of the drawing, various pupils are observing and copying a painting, while in the other vaulted room architects (completing the trilogy painting/sculpture/architecture) are using different instruments, although unfortunately they are not detailed enough to guess their nature, apart from an astrolabe and a compass. On the left, we have two groups of artists (possibly sculptors, to complete the trilogy painting, sculpture and architecture) involved in everyday activities, the group on the left is dissecting and examining a corpse while the group on the right is measuring and studying what is in all probability a live model (or a statue from antiquity).[60]

This scene recalls a description in the fictitious dialogue between the painter Leonardo and the sculptor Phidias in Lomazzo's *Libro de' Sogni*: Phidias states that in order to learn a painter needs "a great deal of drawings, made by excellent men representing girls, old women, youths, old men, and all these sorts of forms, such as the ancients had, together with statues and models (…) and many other things, such as books, medals, anatomies, instruments and the like altogether, without which a painter cannot achieve perfection."[61] To be more precise, the first thing Phidias declares that a painter needs is money, to buy all these sort of things. In any case, this statement appears to be a comprehensive list of the tools shown in the background of the Albertina drawing, with the exception of books, which are, curiously enough, not represented.

While at first glance this drawing might suggest the practice of dissection in the artists' *bottega*, the allegorical context of the representation forces us to reconsider. In this allegorical survey of an artist's career, we have in the background three representations of painting, sculpture and architecture embodied in three fictitious workshops, and we can legitimately doubt the veracity of the practice of dissection as standard.

From a didactic point of view, it is easier to imagine that pupils in the *bottega* had to copy again and again the available drawings and illustrations, as

60 On this drawing's link to anatomical study see also Cazort, Kornell, Roberts *The Ingenious Machine of Nature*, pp. 36–38.

61 Ciardi, *Scritti*, I, p. 92: "Negli studi ci vogliono gli disegni in grandissima quantità da eccellenti e valent'huomini, e de gli vivi, come fanciulle, vecchie, giovani e decrepiti uomini, e di tutte le sorti di forme, come gli antichi avevano, insieme con delle statue o modelli (…) E così molte altre cose che infinite sono ci vogliono, come libri, getti, medaglie, notomie, strumenti e simili, insieme con tutto il resto, senza di quale un non puol riuscir perfetto."

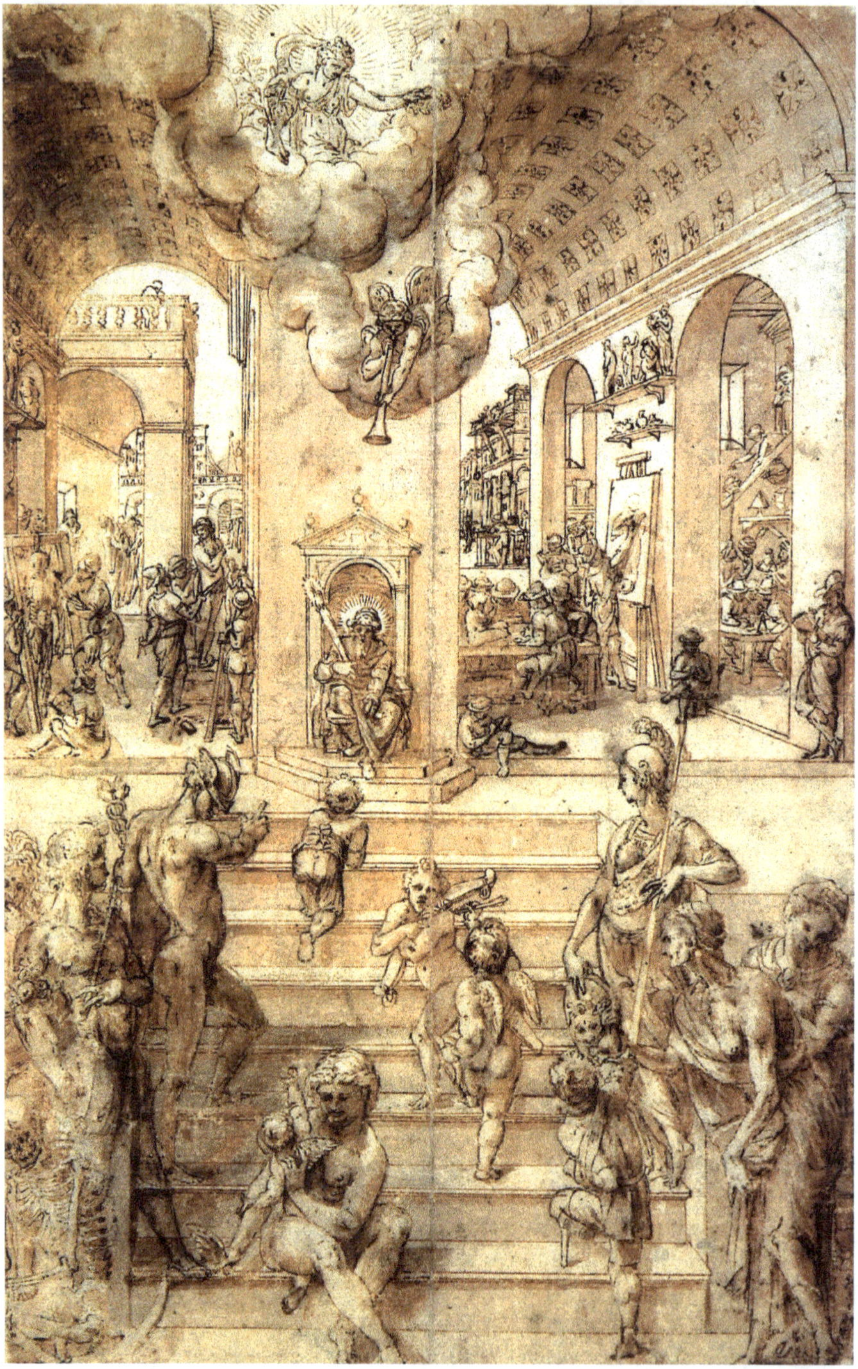

FIGURE 6.14 *Giovanni Paolo Lomazzo (attributed to)*, Allegory of painting or Allegory of the painter's career, *c. 1570, Vienna, Albertina*

Lomazzo suggests, images which will be, later in the seventeenth century, codified in printed editions.[62]

Previous or contemporary scholarly treatises on anatomy may have not been comprehensive for the exercise of painters (such as Jacopo Berengario Da Carpi's *Commentari* -1521) since, although illustrated, they did not provide a sufficiently detailed iconographical apparatus nor were intended specifically for artists to read or to use.

The new visual canon introduced by Leonardo, who failed to publish his *Quaderni d'Anatomia*, but nonetheless intended to (and this demonstrates to some extent a need for such a book), provided painters with an illustrated compendium of anatomical images similar in scope to that of other iconographical books on subjects other than anatomy. These books were circulating in the Italian *botteghe* at the time, along with others such as Cartari's *Imagini degli Dei degli Antichi* (1571) or Ripa's *Iconologia* (1593).[63] The increasing availability of anatomical illustrations most likely encouraged artists to attend dissections and to become acquainted with anatomists –who may have allowed artists to copy specific parts of the bodies that were dissected during the autopsies—but it is unlikely that such practices were done or even discussed in the everyday reality of the artist's workshop, both for practical reasons (lack of time and of the necessary setting) and, not less important, moral reasons. As Katharine Park states, "not all the inhabitants of Renaissance Italy were comfortable with the practice of opening the body."[64] At this point, the copying of illustrations became the essential didactic procedure an artist had to practice in his workshop.

Conclusions

Lomazzo's discourse on anatomy differs in some respects from the treatment that he dedicates to color and perspective. Apart from some scattered chapters and different critical poems in the *Rime*, it is difficult to find anatomical infor-

62 For an discussion on the artists' ways of interacting with anatomists, see A. Hyatt Mayor, "Artists as Anatomists", *The Metropolitan Museum of Art Bulletin*, February 1964, 22.6:201–210.

63 The distinction between the *canone verbale* vs the *canone visivo* introduced by Leonardo is in Domenico Laurenza, "Anatomia e Rappresentazioni Anatomiche tra Arte e Scienza nei Secoli XV e XVI," in *Rappresentare il Corpo*, pp. 31–51.

64 Katharine Park, "The Criminal and Saintly Body," *Renaissance Quarterly*, 1994, *XLVII*:1–33, p. 10.

mation that is not directly linked to the vastly developed subject of proportion, as how to learn "anatomy" (intended as the explanation of how to depict the different parts of the human body), as the author admits, is hard to put into words. Lomazzo therefore decided to include long lists of terms to categorize the parts of the body, many of which have a strong dialectical flavor and were probably taken from the reality of the workshop, and were also influenced by the fictitious language used in the Blenio academy. I have outlined the author's possible connections with Milanese *notomisti* such as the Carcano brothers, considering the possibility that he may have had direct experience of witnessing dissections in the *ospedale maggiore*. Declaring that a painter needs visual material in order to understand how to "make anatomies", Lomazzo incourages painters to copy anatomical illustrations again and again. In this chapter, following the lines of my analysis of color and perspective, I showed how the artistic context of Milan influenced Lomazzo's suggestions in his book. I presented some anatomical drawings derived from Leonardo and attributed to the colleague of Lomazzo and member of Blenio Annibale Fontana that were circulating in the city at the time, as an example of which kind of visual materials the author refers to and in which ways this material was shared among artists.

This series of drawings attributed to Fontana testifies to the strong influence of Leonardo, more than half a century after his death, on Milanese artists' working methods and especially their practical didactic approach to the study of the body. As Carmen Bambach in her book *Drawing and Painting in the Italian Renaissance Workshop* brilliantly summed up: "As unique documents of the gestation of works of art, drawings in particular hold the promise of a glimpse into the private wold of genius. But basic questions about many significant Italian artists still await detailed answers: How did they draw? What were the purposes of their drawings? And what do their drawings reveal about their general design methods?"[65]

The drawings attributed to Fontana and the others presented in this chapter give us a glimpse of the widespread practice of copying as a way of sharing anatomical knowledge. As I argued, this practice became gradually more public in Lomazzo's time thanks to the establishment of informal places for artists outside the painter's workshop, like the *Accademia di Blenio* may have been.

65 Carmen Bambach, *Drawing and Painting in the Italian Renaissance Workshop. Theory and Practice, 1300–1600* (Cambridge: Cambridge University Press, 1999), p. 11.

General Conclusions

Lomazzo constitutes a crucial case study of learned artist and writer on art who embeds in his writings not only the theoretical discourse on the different parts of paintings, but ideas and information taken from the practice of his art. The topics that he discusses in his books and prominently in the *Trattato dell'Arte della Pittura* reflect the appropriation of earlier sources (writings on art, classical authors, philosophical works and collections of workshop's material), and at the same time they are indicative of the author's direct involvement in the cultural and pictorial environments of the city of Milan. His contemporaries recognized his position as painter, academician and writer, and when examining closer the community of artists and men of letters of which he was part, we get a clearer understanding of where the material that he decided to include in his writings comes from. In his books we come across topics that are a reflection of the discussions he was aware of through direct contact with different people, both in and outside the city. The sources that he explicitly quotes in the *Trattato*, building his long lists of names of artists, and classical authors, are sometimes misleading and cannot legitimately be the only ones from which he drew information. It has been noted by different scholars in the past how Lomazzo's material was "overloaded with antiquarian erudition" in all his parts, especially in the *Trattato* and in the *Idea*.[1] This is certainly true, however, this study outlined how his philological interest in previous sources is combined with the artist's awareness of contemporary debates (such as on perspective and optics) and with his close link to the reality of the workshop. It is especially true for the terminology that he uses to classify the different parts of painting.

In the first part of the research, I gave an overview on the artistic environment of Milan in the second half of the sixteenth century and I outlined some of Lomazzo's places of interaction, focusing on the painter's connections not only with artists but with men of letter and analysing in detail the *Accademia de la Val di Blenio*. I subsequently presented the figure the patron of the Blenio academy, Pirro Visconti Borromeo, and the project for his Villa in Lainate.

In the second part, I focused on three main aspects of the painter's theoretical discussion: color, perspective and anatomy. In the chapter on color, I analysed Lomazzo's use of sources such as earlier collections of recipes, which he decided to include in his chapter on the theory of colors, and how these more technical parts fit into his general discourse on the subject. In the part on perspective, I underlined the mixture of previous discourses by authors who

1 See for instance Barasch, *Light and Color*, p. 136.

 | DOI 10.1163/9789004330269_010

discussed on art with those discussing contemporary optical theories, the interest of the author in a mathematical construction of perspective as well as in the practical methods and instruments to use. Regarding the part on anatomy, I have interpreted Lomazzo's scattered discourse as a reflection of the difficulty to talk about "how to learn anatomy" with words and without of use of drawings as didactic material, presenting the *lessico* that the painter uses to define the different parts of the body in comparison with the terminology used in the Codex Huygens, finally analyzing a series of understudied drawings belonging to his circle of artists.

The heterogeneity of the artist's interests and influences is well reflected in his dealings with fellow members of the academy of Blenio. Lomazzo uses both his literary skills and his connections to erudite men to grasp or at least to get an idea of different bodies of knowledge, such as optical debates and astrological theories, in order to build his complex system of painting. Beyond the strict categorization that he achieves, and beyond the fact that the material he presents was not always coherently organized in printed version because of the aforementioned editorial difficulties, the analysis of the topics discussed allowed us to realize how the subjects Lomazzo presented intermingle with one another. We found references to:

Light in the part on Color and Perspective
Color and Light in the part of Perspective
Perspective in the part on Color.[2]

All subjects are interconnected with each other and they present sources related to artistic practice, "scientific" theories, as well as artistic and classical literature.

We have encountered particular difficulty in reconciling the concepts of theory and practice in Lomazzo's writings, as the technical information is often misplaced in the text, at least if one follows the author's misleading categorization. Lomazzo's will to give us a glimpse of the reality of the workshop placing practical information among his vast theoretical discussion on painting goes into the direction of those books which Claire Farago calls "compendia of

2 I partially disagree with Moshe Barasch, who in his analysis of Lomazzo's colors states that color in the author's discourse is independent of light. (See Barasch, *Light and Color*, p. 159 ff.). Namely this is actually so, because, as he states, Lomazzo is the first writer on art to make separate chapters on light and on color, but on the content level we often find different hints to light and the effects thereof in Lomazzo *Del Colore* as well as in *Della Pratica* and in the later *Idea*.

information to be utilized by practicing artists".[3] However, it does not limit itself to that: looking beyond the declared structure of the *Trattato* and focusing on the actual information given in each part, we could infer that the artist's writings, though they should still be inscribed in the tradition of Italian *trattati* on art, have also something in common with more technical books, such as the "how-to" books, writings which embedded practical or rather information on how to do or make something (as for instance recipes for pigments).[4]

It became clear how a growing number of Milanese artists were inclined to gather and interact beyond the reality of the *bottega* and beyond the *Scuola di San Luca*. Overall, the establishment of the *Accademia di Blenio* can be seen as a first attempt to build an informal place of exchange of ideas and opinions on the state of the arts, a place for artists who knew "how to read and write" to share their literary achievements and their passion for burlesque poetry. The academy had a strong regionalist component with its roots deeply nestled in the city in which it was established. The members promoted themselves as "true lovers of all the letters and arts", while at the same time put forward an alternative canon to the Florentine tradition of academies and, implicitly, of art. Though it would be misleading to state that this academy was a legitimate forerunner of Federico Borromeo's academy of art, as the two had little in common when looking at their structures and objectives, its creation and its popularity among Lomazzo's colleagues testifies to the need for such a place, in a way similar to how his treatise testifies to the growing interest of Milanese artists in writing and publishing books not only on the theory, but also on the practice of painting. As is well known, other books on art would follow the *Trattato* in the next few years, most notably Gregorio Comanini's *Il Figino*

3 See Farago, *The Fabrication*, p. 9: "New forms of artisanal knowledge ordered to practice did not privilege individual authorship – both the Codex Huygens and the abridged version of Leonardo's treatise on painting circulating at this time functioned as compendia of information to be utilized by practicing artists. Yet at the same time, new appreciation for individual authorship, driven partly by the economic opportunities that print technology offered, put pressure on what had previously been collectively transmitted as workshop 'secrets.'"

4 Research on how-to books (especially but not only recipe books) is being carried out by already mentioned scholars such as Pamela Smith, Pamela O' Long, Sven Dupré and Elaine Leong. See for instance Elaine Leong and Alisha Rankin (eds.), *Secrets and Knowledge in Medicine and Science*, 1500–180 (Farnham (Surrey): Ashgate, 2011); O. Long, *Openness, Secrecy, Authorship: Technical Arts and the Culture of Knowledge from Antiquity to the Renaissance* (Baltimore: The Johns Hopkins University Press, 2001); Smith, *Making Knowledge in Early Modern Europe: Practices, Objects, and Texts, 1400–1800* (Chicago: University of Chicago Press, 2008).

overo del Fine della Pittura, named after the most renowned of Lomazzo's pupils, Ambrogio Figino.[5]

Regarding the fate of the *Trattato*, I hinted at it in the introduction but ultimately decided not to include the scattered evidence available as part of this study, as it proved not substantial enough to pin down the circulation of Lomazzo's treatise within Milan and beyond. Still too little is known about the actual owners and readers. We know that artists of the same and of the following generations, mostly connected to the Milanese context, such as Daniele Crespi, Giulio Cesare Procaccini, Pellegrino Tibaldi, and Jacopo Ligozzi owned a copy of the Treatise. In some cases, it was even donated by the author himself, who was without doubt a good self-promoter.[6] This knowledge relies mainly on inventories and exchanges of letters. Among other colleagues of the artist, as we mentioned earlier we know that the painter Leone Leoni from Arezzo (1509–1590) knew of it. In a significant letter from Bologna by the unknown Valerio Angelini, dated 29 October 1586, and published in appendix to Lomazzo's *Rime* (1587) is written that:

> I hesitated until now to present myself (as I could not do otherwise) through these lines to you, not to your eyes because of your unfortunate destiny, but to your most noble and sharp spirit, and I wish in my name and in others, as we imitate your merit, to pay our due respects. Many are the reasons for this, which would be long and boring to explain. But what moves me to do this now is the authority of the great *cavalier* Leone Leoni, who praised your name to me most rightfully, and made me think that this statement of my opinion and regard for you would not come unwelcomed.[7]

---

5 Comanini, *Il Figino* (op. cit).

6 See Chai, *Idea*, p. 204, n 131, and Agosti and Stoppa (eds.), *Bernardino Luini e i suoi Figli*, p. 301. On the topic of Lomazzo's self-promotion I wish to mention the on-going research by the PhD candidate Anna Sgobbi at the Ludvig-Maximilians-Universität in Munich.

7 Lomazzo, *Rime*, Letter of Angelini published as appendix: "Ho differito fino a questo tempo a presentarmi (poichè altrimenti non posso) col mezzo di queste righe al cospetto, se non degli occhi (colpa della fortuna) almeno del chiarissimo e lucidissimo spirito di V.S. e in nome mio, e d'ogni altro che, come io, emuli il valor suo, a renderle il debito ossequio, e tributo. Le cause di ciò sono molte, e sarebbono lunghe e noiose; ma quello, che mi muove ora secondariamente, si è l'autorità dell'onoratissimo, e lodatissimo, e da me riverito sig. cavalier Lione Lioni, il quale celebrando con infinita ragione meco particolarmente sempre il suo nome, ha fatto, ch'io sono entrato in pensiero, che a V.S. non sia per essere ingrata questa (qual ella si sia) significazione dell'opinione, e dell'affetto mio."

As just mentioned above, Pellegrino Tibaldi also called Pellegrino Pellegrini (1527–1596) owned a copy of the Treatise and of the Idea in his library. The inventory from 1600 registers the two books in double copy (*doppia copia*), among the obvious books mentioned before, such as Cartari's *Imagini de gli Dei degli Antichi* and Boccaccio's *Genealogiae deorum*.[8] As Buratti Mazzotta pointed out, the arrangement of themes in Pellegrini's book *Architettura* and in Lomazzo's treatise is similar in some ways. It is therefore quite plausible that the former read the *Trattato dell'Arte*.[9]

Domenico Zampieri called Domenichino (1581–1614) may have owned a copy of the treatise, but the evidence is far from satisfying. As Bialostocky pointed out, among the books owned by his pupil Francesco Raspantino (1664), who inherited his studio and some of Domenichino's books, there is a copy of an unspecified *Trattato della Pittura*. Bialostocky suggests that it may have been Lomazzo's, although the lack of the author's name is indeed an obstacle to verify it.[10] In general, we see how it is predictably less difficult to find later sources mentioning our author among the "authorities" that a painter needs to know to learn the art of painting, on which we will mention only some. In 1674 the painter and artist biographer Luigi Scaramuccia, in *Le Finezze dei Pennelli Italiani*, mentions Lomazzo at the very end of his book, in his *Catalogo degl'autori che hanno scritto di pittura promesso al cap. 27, p. 79*.[11] Curiously, he mentions both the *Trattato* and the *Forma delle Muse*, but not the *Idea*. Other authors who mention Lomazzo in the seventeenth and eighteenth centuries include Pellegrino Antonio Orlandi, or, outside of Italy, John Dryden.[12] This evidence, however important, cannot be considered sufficient

---

8 Marzia Giuliani, "La Biblioteca di casa Pellegrini," *Studia Borromaica*, 1998, *XII*:39–100, p. 42. I thank Professor Giulio Bora for the suggestion.

9 See Pellegrino Pellegrini, *L'Architettura*, ed. by Buratti Mazzotta, Milan 1990, especially p. LII, quoted in Giuliani, "La Biblioteca di casa Pellegrini", p. 46.

10 Bialostocki, "Doctus Artifex," p. 14.

11 Luigi Scaramuccia, *Le Finezze dei Pennelli Italiani*, p. 217–219.

12 John Dryden, who probably had knowledge of Lomazzo through Haydocke's translation, lists his books among the technical treatises, along with other authors such as Franciscus Junius and Leonardo. See Sir Walter Scott (ed. by), *The Works of John Dryden* (Edimburgh: Hurt, Robinson and Co., 1821), 18 vols., vol. 17, pp. 415–416: "Amongst others, Leonardo da Vinci (though without method); Paulo Lomazzo, whose book is good for the most part but whose discourse is too diffusive and very tiresome; John Baptiste Armenini, Franciscus Junius and Monsieur de Cambray, to whose preface I rather invite you, than to his book." In Bologna the writer Pellegrino Antonio Orlandi mentions Lomazzo in the *Abcdario pittorico*, at the end of the first part, explicitly saying that he used these modern authors

in order to understand the extent to which the *Trattato* was considered of interest by painters in Lomazzo's own time. His books ultimately aimed at real or aspiring art lovers, as one can more clearly see when considering reception of the artist's theoretical discourse outside of Italy, especially thanks to the translation into English done by the scholar Richard Haydocke only fourteen years after the first edition of the book. How Haydocke came across the book and his reason for translating it, is still a matter of speculation, but it allows us to think that the book gained a certain fame in Italy during the time of the scholar's visit to Italy, as he must have come across the treatise and decided that it was interesting enough to translate its first five books -while the practice of painting and the book on the *historie* were significantly not taken into consideration.[13]

Analysing both the spaces of interactions such as the artists' academy and some of Lomazzo's sources of information as well as the terminology the he decided to use for the part on color, perspective and anatomy, the research opened up new questions and threads, such as the ones related to the education of artists and the transmission of pictorial knowledge from the painter's workshop into treatise's form. For what concerns the artist's education, we have seen how Lomazzo conveys many times the notion of the learned painter. This declaration that the painter to be good had to be also "literary competent" has usually been ascribed within the artist's attempt to raise his social status, like the growing involvement of artists in literary academis has been usually interpreted as an attempt to obtain more commissions and to extend the range of their clients. However, Lomazzo was probably not only trying to promote himself and the art from his region, but to establish a legitimate new "canon of the good artist" and of the good art, gathering on one side all the previous knowledge on art that he could have access to (often omitting to quote the

---

to write this book. Lomazzo is mentioned both as a painter (*Tavola prima, Cognomi dei pittori*, p. 436) and as a writer in the *Tavola II degli Autori e scrittori*, p. 458. Here all his works are mentioned (*Trattato, Rime, Idea, Della Forma dell Muse*). He is described as a good painter and teacher (Pellegrino Antonio Orlandi, *Abcdario Pittorico* (Bologna: Costantino Pisarri, 1704).

13 It is plausible to infer that Haydocke must have thought that these "practical" books, which he decided not to translate, were more of interest for practitioners than for art lovers. On the reception of Lomazzo in other European countries much research still needs to be done. For the reception of his Treatise in France, I wish to mention the ongoing research of Dr. Stéphanie Trouvé and her recent presentation "Lomazzo and France: Hilaire Pader's Translation; Theoretical and Artistic Issues" (Renaissance Society of America, Boston 2016, available online at: <https://hal.archives-ouvertes.fr/hal-01319315 v3/document>).

source where he took information from) and on the other side using and embedding in his writings the terminology of the workshop and his expertise as a painter.

We focused on the *Trattato* (and to some extent on the *Idea*), underlining the importance of the context in order to understand the content of this book. However, other writings should be taken into consideration concerning the artist's opinions. I have shown how the *Rabisch* and the *Rime* especially, which were traditionally separated form the artist's "official production" and confined under the label of "facetious writings", constitute an essential source in order to pin down with more certainty Lomazzo's judgements on and beyond painting. They give useful hints on a variety of topics: the artist's critic and praise of bad and good art, his position towards alchemists and astrologers, and, most importantly, his connections with artists and learned men. These books therefore play a crucial role in order to get a better understanding of the artist's life and intellectual achievements, as well as of the material he collected and presented in the *Trattato*.

APPENDICES

APPENDIX 1

Contract between Giovanni Paolo Lomazzo and Giulio Claro, *Reggente* in Milan, dated 1561

MDLXIX, Sabato a di nove del mese di luglio, in Milano.
Convengono per tenore del presente scritto, il molto illustrissimo Signor Giulio Claro Reggente di Milano per una parte, e il nobile Messer Gio. Paolo Lomazzo pittore, figlio emancipato del nobile Messer Gio. Antonio, come appare per istrumento rogato per M. Giov. Pietro Carcano, nodaro di Milano, l'anno 1561 prossimo passato, Porta Ticinense, parocchia Santa Maria Beltrade, per l'altra parte.

18 Oct. 1561:
DOMINI IO. ANTONIUS DE LOMATIO FILIUS QU. DOMINI GIORGII, PORTAE TICINIENSIS, PAR. S. MARIAE BELTRADIS ET DOMINUS IO. PAULUS, EIUS DOMINI IO. ANTONII FILIUS, MAIOR ANNIS VIGINTI DUOBUS UT DIXI, ET ETIAM AC EIUS ASPECTU CORPORIS EVIDENTER APPARET.

Primo ch'el detto Signor Claro sia tenuto dare, come sin ad ora dal detto Messer Giovanni Paolo presente, e qual confessa aver ricevuto dal Signor reggente ottanta disegni di mano de' pittori famosi, e cinque quadri, uno di mano di Tiziano, uno di mano di Michelangelo, uno di mano del Luino, uno di mano di Bramantino e uno di mano del Soiaro, e di più molte altre cose del camerino o studiolo del detto Signor Claro, quali ascendono al valore di scudi cento cinquanta d'oro, secondo che fra le dette parti s'è convenuto, e detto Messer Giovanni Paolo ha protestato e protesta. Secondo convengono ch'al detto Messer Giovanni Paolo sia tenuto, e così promette, in cambio delle dette cose, e per pagamento del detto prezzo, consegnare al detto Signor Claro, o suoi agenti qui in Milano quattordici pitture fatte di sua mano della manera infrascritta e nelli termini infrascritti, fatte e finite a tutte sue spese. E primo promette per tutto il mese di Settembre, un estratto del Cristo del sepolcro a Milano, della medesima grandezza che è quello del detto Sepolcro affresco di Bramantino, copiato dai contemporanei, all'Ambrosiana, e con l'istesse figure e colori, qual s'estima scudi venti.

Item promette da qui a Natale prossimo avvenire, uno quadro d'un Cristo assettato su la Croce, innanzi alla crucifissione, con gli crucifissori attorno e altre figure opportune, nell'atto che disse *Pater ignosce illi, etc.* e questo quadro sia d'uno braccio per lato almeno, e s'estima d'accordo scudi dieci.

Ultimamente promette dodici ritratti de' dodici re Barbari vestiti e armati nel modo e forma che li descrive Vuolfango Lazio nel suo libro intitolato *De Migrationibus*

 | DOI 10.1163/9789004330269_011

Gentium intieri e grandi più del naturale, e con il suo spazio per due righe e lettere che dirà esso Signor Claro, questi fra un anno e meggio, da poi che detto Signor Claro gli averà consegnato detto libro di Vuolfango, quali s'estimano scudi dieci l'uno sottosopra.

E nelle sopradette cose, promette detto Messer Gio. Paolo usare ogni diligenza a lui possibile, così nelle figure come nel colorire senza risparmiare spesa, danno e interessi in caso di contravvenzione, volendo che il presente scritto abbia forza di qual si voglia più gagliardo istromento, qual sopra ciò si possa fare, e con patto espresso che possa esser convenuto in forma di camera, ecc., e con rinunzie e promesse debite e giuramento, ecc., e in fede hanno fatto scrivere la presente convenzione, e sottoscrivere dalli sottoscritti testimoni.

In Milano a dì soprascritto

Io Giulio Claro Reggente di Milano affermo quanto di sopra si contiene
Io Gio. Paolo Lomazzo prometto e giuro come di sopra
Io Gio. Antonio Homacino ho scritto e sottoscritto per testimonia de volontà delle parti
Io Jeronimo Maderno fui presente per testimonio de consentimento delle parti

ANNOTAZIONE:
Io. Sudetto Gio. Homacino procuratore speciale del suddetto molto Illustrissimo Signor Reggente Claro, confesso aver avuto dal detto Messer Gio. Paolo li detti duoi quadri, cioè il Cristo alla forma di quella sopra la porta di Santo Sepolcro di Milano, e l'altro Cristo assentato sopra la Croce, come dice nel scritto, fatto in pietra, e in fede.

In Milano a dì XII Agosto 1572
Io. Gio. Antonio ho scritto e sottoscritto

APPENDIX 2

L'interogaçiglion ch'o s'han da fa dar gran Scanscieré pos ra gneregada a col ch'o vùr intrò in dra Vall de Bregn

Com' vaga ligò or fegn.
Come vaga ra còrda e 'd che longhezza.
Come va ra fusella.
Com' va ra brenta.
Com' va or pagliù.
Come vaga consciò or pagliù in dra brenta, quand o rè drent or vign.
Come besògna ch'o siglia or vign ross bogn.
Come zògna ch'o siglia or vign blanch a ess de col bogn.
Come va or sciatt.
Come vaga or spontogn afferò.
Come va i sciavatt.
Come va or sacch da portò in spall e in cò, e com'vaga or scossarign da mett drent or sciatt.
Com' va consciad or sacch quand o's vur portò ra carga.
Come vaga i cortigl' da scortegò i cavritt.
Com' se scortega i cavritt. Com' o's faga a tù ra pell al cavrett.

Or fign degl'interogaçiglion.

 | DOI 10.1163/9789004330269_012

APPENDIX 3

Difinicione della tavola sopra detta

Al fine, per darti più chiaramente a intendere questa lingua, con la quale tu potrai dire tutto quello che ti verrà in pensiero, e per far questo, pigliarai li vocaboli, over parole, primamente che finiscono in queste cadenza cioè io an, en, in, on, un, à quali in questa nostra lingua gli farai all'ultimo un, g, & un,n. Si come per esempio, del primo si dirà, mano: va detto magn, e nel secondo, seno, va detto segn; e nel terzo, vino, va detto vign; nel quarto, dono, va detto dogn; nell'ultimo, uno, va detto ugn, et ancora nei medesimi vocaboli che finiscono per due, n, gli porrai gli'istessi, g e n, si come per esempio si vede, circa il primo anno, va detto agn, e nel secondo, denno, va detto degn, e cosi si può seguitare, in quante parole possono mai cadere in tali desinenze. E ancora si ha d'avvertire che in quanti vocaboli dove erano al mezo ò al fine, di loro due vocaboli appresso, se gli ha da fare un, g&unl, in mezo di quelli, come farebbe per essempio di tutti gli altri. Dio, o Dee, va detto Digli & Deglie. Oltra di ciò, si usa in questa ancora per levare quelle due vocali al principio, di dire in vece di Giuliano, Slurigliagn, et di Aurelio, Sluregligl', così si va variando. Ma tornando al s, di Bregno questa, s. serve ancora a levar via, a tutti li vocaboli dove entrano due c, farli questa s, et un c, come per essempio si vede, in cacciò, va detto casciò, et ancora questa s serve nei vocaboli, dove entrano i c, facendoli questa s inanzi, come si trova per esempio in ciancia, over pancia, va detto scianscia, et panscia, et serva ancoa questa s in levar via el, g, et le, si come per esempio si vede in ingegno, va detto insciegn, et tutti li vocaboli dove entrano due t, in loco di quelli vanno posti due c, et un h, come si vede per essempio in scritto, overo letto; si ha da dire scricch, lecc, et parimente si farà in alcuni altri, che habbiano un solo t, et porli un c et un h, et ancora a quanti t entrano nella penultima lettera del vocabolo, si ha da porgli un d, un a e un o, come sarebbe per essempio degli altri coronati, va detto cornad, coronà e coronò. Ma bisogna avvertire che nel c, che entra in loco del t, di non fare errore, perché in vece di parlar di Bregno, si parlaria Bergamasco, o da Villano, et ancora in molti t, che entrano per diversi nomi, i quali per tali t, si servono ancora in questa, come carte, marte, arte, et simili che si dicono cart, mart, art et tutti quei vocaboli dove entrano o, va portato quel o in u, come si vede in huomo, pomo, va detto hum o pum, et ancora tutti i, l, vanno cangiati in tanti r, come sarebbe a dire lo mio bene, va detto or me begn, et ancora serve, il medesimo, o per i, e per e, si come si vede nella tavola sopra detta con molte altre parti, che sarebbono da dire bastando, solamente havere accennato il fondamento, solo di questa lingua, la quale dipende fora, ma rozzamente dalla lingua Toscana, ign dra qual per diu o vero g va tanta consideraciglion are à cogl più che a mi, quand ghe vedeva, or trovà gl'invenciglion degl insturigl', e tutt col co se ha da met in pinchiura, e co sto sign, om recomanda tucch

 | DOI 10.1163/9789004330269_013

cogl' coglign de ra Vallada de ra qual scinsciand, di mi è di su fidil soghitt, o spe prest de da fura gragn quantitad di sversarigl de divers lenguag à honò de costa grandissima Vallada de Bregn.

Or fign de r'upra.

APPENDIX 4

Straducc dra vall de Bregn

Che or Nabad facc e incoronad par magn dor gran Scanscieré cor consentiment dor Consegl' siglia perpetugl, salvv se par soa colpa no 'r meritass da ess levad dra abadiglia par sti tre casogn:

Rebiglion dra Vall.

Non vorè ascoltà or Consegl' or dì de Consgl.'

No vorè fa giustiçiglia.

Che onzugn no's possa scietò in dra Vall se no r'è admiss dar Nabad e interogad dar gran Scancieré di resogn de sòtt.

Ch'o s'abiglia a obedì or Compà Nabad e 'd fa tutt ch' o' gh comandarà par Bregn.

Ch'o 's mostra tutt col ch'o's sa fa con ra penna in magn da col ch'o vul intrà in tra vall.

Ch'o no s'ascieta onzugn s'or no n'è virtuglios in qualcoòssa, e principalment in degl' art liberal.

Che gl'interogaçigliogn o's fagan pos ra gneregada ch'o farà corù a tucc i savigl de Bregn cor Nabad.

Che pos l'interogaçigliogn or Nabad r'ascietta indra Vall e ch'o's lassa basà ra magn.

Or fign degl'straducc.

 | DOI 10.1163/9789004330269_014

APPENDIX 5

Inventory 24th January 1604, doc. B, notary Benedetto Coerezio, f. 20578

Uno scrittorio di avorio bello con figurine d'argento et fresi d'argento

Un quadrettino grande come un offitiolo di una madonna col puttino in braccio e S. Giuseppe, e fride di avorio n.1

Una borsa a gucchia di seta n.1

Medaglie di metallo vecchio n.12

Una cinta di seta nera n.1

Un altro scrittorio quadro di ebano e lavori intarsiati di osso bianco er è quadro e una figulina d'argento nel mezzo n.1

Un quadrettino picciolo d'ebano e la morte disegnata n.1

Una crosetta piccola d'osso bianco e dentro alcune Reliquie n.1

Medaglie d'argento bone n.5

Altre Medaglie d'argento n.5

Un altro scrittorio d'avorio e le corniciette di osso bianco

Due tazze di paglia d'India adorate per dentro n.2

Fibie d'argento per un balandrano di diverse sorte n. 36

(...)

Un'aquila d'oro che magna un corpo parimenti d'oro con sotto una preda nera di aspiri n.1

Horologi da soli diverse fatte n.6

Un quadrettino nero con un alla depinta d'Alberto Duro e sopra il cristallo n.1

Un altro stucchietto e dentro li ferri di ottone per disegnare n.1

Una tazza di diaspidi orientale con (...) e piede d'oro n.1

Una navicella di diaspidi orientale ed due fresi d'oro al piede

Una Tazza col suo piede di lapis armenio sopra l'humour malinconico n.1

Una Tazza d'argento sopra dorato e nel mezzo una preda di bazar occidentale n.1

(...)

Uno scrittorio di avorio bello con figurine d'argento et fresi d'argento

Un quadrettino grande come un offitiolo di una madonna col puttino in braccio e S. Giuseppe, e fride d avorio n.1

Una borsa a gucchia di seta n.1

Medaglie di metallo vecchio n.12

Una cinta di seta nera n.1

 | DOI 10.1163/9789004330269_015

Un altro scrittorio quadro di ebano e lavori intarsiati di osso bianco er quadro e una figurina d'argento nel mezzo n.1
Un quadrettino picciolo d'ebano e la morte disegnata n.1
Una crosetta piccola d'osso bianco e dentro alcune Reliquie n.1
Medaglie d'argento bone n.5
Altre Medaglie d'argento n.5
Un altro scrittorio d'avorio e le corniciette di osso bianco
Due tazze di paglia d'India adornate per dentro n.2
Fibie d'argento per un balandrano di diverse sorte n. 36
(...)

APPENDIX 6

Inventory, 11th November 1611, Fondo Litta, carte 32

Notta de quadri, tavole e altre cose quali si lasciano per ornamento della Fontana

Uno orologio con scrittorio d'ebano con guarnizione d'argento con sua cassa di veluto turchino

Una fontana a monticello con coralli

(...)

Una tavola di pietra interciata con pietre di diversi colori

Due altre tavole di pietra mischie

Quadri diversi

Tavola di pietra interciata ut supera minor dell'altra

Un cornetto d'animale

Un pezzo di vite del tempo di Noé

Una ventalina turchesca sopra dorata

Due fonghi impietriti

Un pezzo di palma indiana con lettere sopra

Diaspro orientale legato il pie' in oro

Madreperla legata in argento

Tazza di lapis armeno per l'umore malenconico

Madre di perla legata in argento

Mazzo di noce d'India legazo in argento

Un calamaro atornito in ovato

Una coppetta indiana miniata d'oro

Due lumaghe di madreperla miniate con oro

Due tazze di paglia all'indiana dorate

(....)

 | DOI 10.1163/9789004330269_016

APPENDIX 7

Libro III *Del Colore* (1584)

i. Della virtù del colorire
ii. Della necessità del colorire
iii. Che cosa sia il colore e le sue spezie e d'onde si cagionino i colori
iv. Quali siano le materie nelle quali si trovino i colori
v. Quali colori a ciascuna spezie di dipingere si confacciano
vi. Delle amicizie et inimicizie de colori naturali
vii. Quali colori e meschie faccino l'un colore con l'altro
viii. Della convenienza c'hanno fra loro i colori chiari et oscuri
ix. De i colori trasparenti e come si adoprano
x. Dell'ordine che si tiene in fare i cangianti
xi. De gl'effetti che causano i colori
xii. Del colore nero
xiii. Del colore bianco
xiv. Del colore rosso
xv. Del color pavonazzo
xvi. Del colore giallo
xvii. Del colore verde
xviii. Del colore turchino
xix. Di alcuni altri colori

 | DOI 10.1163/9789004330269_017

a. Paduan Manuscript (Merrifield, pp. 648–717), *Ricette per fare ogni sorte di colori* (Chap. I – *De colori in generale, e di quali materie si componghino*)

b. Lomazzo, IV chapter of III book of the *Trattato*, *Quali siano le materie nelle quali si trovino i colori*

a. Il bianco si fa col gesso, calcina, marmo pesto, gusci d'ovo bene polverizati, et settazzati, e con osso di sepia benissimo macinato.

b. Tra i colori materiali che si usano generalmente a questi tempi se ne ha cognizione di molti i quali tutti hanno i suoi particolari colori. E prima quelli che fanno il bianco sono il gesso, la biacca, il bianco e il marmo trito. Evvi ancora un'altra cosa che a fresco fa restare i colori nel modo che si dipinge quando la calce è fresca, e questa è una delle rare invenzioni che sia nella pratica dell'arte, cioè il guscio delle uova tritato minutamente e con quello mescolare tutti i colori più e meno, secondo se gli appartengono. E il bianco che non si può sfiorare, tritato minutamente, è buono a colorire le carni perfettamente in fresco.

a. Gialdo si fa col gialdolino di fornace di Fiandra, et d'Alemagna, orpimento e ocra, col zaffarano et gomma gute, quali sono per l'acquerello.

b. I colori che fanno il giallo sono il giacobino di fornace di Fiandra e di Alamagna, e l'oropimento scuro e l'ocra.

a. Turchino si prepara con gl'azuri oltramarini et ongari et altri si fa ancora con gli smalti, e smaltini d'ogni sorte, massime con quelli di Fiandra, che sono li migliori, con li Biadetti e simili.

b. Quelli che fanno il turchino sono gli azzurri oltremarino, l'onagro e gli altri, e ancora gli smalti, come quello di Fiandra che è il migliore de gl'altri tutti.

a. Verde si fa con li verde azurri, verderami, verdetto, che si chiama gialdo santo, e tira al gialdo, terra verde, verde porro, e poi l'acquerella, verde di vescica, o sia pasta verde, succo di ruta, di gigli azurri.

b. Quelli che fanno il verde sono i verdi azzurri, il verderame, il verdetto, che si chiama santo, ma tira al giallo, e ancora la terra verde, i verde di barildo.

 | DOI 10.1163/9789004330269_018

a. Morello si fa con morello di ferro, e di sale, vitriolo romano cotto, et indico oscuro, e per acquarella il tornasole.

b. Il morello di ferro e quello di sale fanno il morello et oltre di ciò il vetriolo cotto, il cilesto e l'endice oscuro.

a. Rosso si fa col cinaprio, terra rossa, lacca fina. Ranzetto si fa col minio et oropimento arso.

b. Quelli che fanno il rosso sono i due cinabri, cioè di minera et artificiale, e la terra rossa, detta maiolica. Il rosso sanguineo lo fanno le lacche tute, e il ronzato lo fa il minio et ancora l'orpimento arso, il quale si dice color d'oro. E questa è l'alchimia de i pittori veneziani.

a. Negro si fa col fumo d'oglio di noce arso, guscio di mandola arso, fumo di ragia, et terra negra.

b. Finalmente, quelli che fanno il nero sono l'avorio arso, il guscio della mandola, il nero di balla, il fumo di ragia e finalmente il nero di scaglia detto della terra nera.

a. Umbra delle carni si fa con terra d'ombra verde arsa, e spalato.

b. Ombra delle carni oscura è fatta dalla terra di campana, dalla terra d'ombra, detta falzalo, dalla terra verde arsa, dallo spalto, dalla mummia e da altri simili.

Bibliography

Ackerman, Gerald Martin, *The Structure of Lomazzo's Treatise on Painting*, Doctoral Dissertation, Princeton University, 1964

Ackerman, Gerald Martin, "Lomazzo's Treatise on Painting", *The Art Bullettin*, XLXI, 4, December 1967, pp. 317–326

Ackerman, James S., "Leonardo's Eye", *Journal of the Warburg and Courtauld Institutes*, XLI, 1978, pp. 108–146

Ackerman, James, *Distance Points: Essays in Theory and Renaissance Art and Architecture*, Cambridge (MA) 1991

Agosti, Barbara, "Federico Borromeo e Girolamo Borsieri: Un Capitolo Lombardo di 'Fortuna dei Primitivi'", *Rivista Archeologica dell'Antica Provincia e Diocesi di Como*, CLXXIV, 1992, pp. 225–240

Agosti, Barbara, "Contributo su Annibale Fontana", *Prospettiva*, LXXVIII, 1995, pp. 70–74

Agosti, Barbara, "La Pinacoteca Ambrosiana", *Prospettiva*, LXXXVII–LXXXVIII, 1997, pp. 175–181

Agosti, Giovanni; Stoppa, Iacopo and Sacchi, Rossana (eds.), *Bernardino Luini e i suoi Figli*, Officina Libraria, Milan 2014

Aimi, Antonio; De Michele, Vincenzo and Morandotti, Alessandro, *Musaeum Septalianum. Una Collezione Scientifica nella Milano del Seicento*, Museo Civico di Storia Naturale, Milan 1984

Albinico, Simone, *Biografie degli Artisti dell'Accademia della Val di Blenio*, Skira, Milan 1998

Alexander-Skipnes, Ingrid (ed. by), *Cultural Exchange between the Low Countries and Italy (1400–1600)*, Brepols, Turnhout 2007

Ames-Lewis, Francis, *The Intellectual Life of the Early Renaissance Artist*, Yale University Press, London and New Haven 2000

Antal, Frederick, *La Pittura Fiorentina e il Suo Ambiente Sociale nel Trecento e nel Primo Quattrocento*, Einaudi, Turin 1960

Aristotle, *De anima libri III: quibus accedunt libri De sensu et sensili, De memoria, De somno, De somniis, De longitudine et brevitate vitae, De iuventute et senectute, De vita et morte, De respiratione/ad optimorum librorum fidem accurate editi*, Sumtibus et Typis Car. Tauchnitii, Leipzig 1831

Armenini, Giovanni Battista, *Dei Veri Precetti della Pittura*, F. Tebaldini, Ravenna 1587

Azzolini, Monica, *The Duke and The Stars: Astrology and Politics in Renaissance Milan*, Harvard University Press, Cambridge (Mass.) 2012

Baldinucci, Filippo, *Notizie de' Professori del Disegno*, D.M. Nanni, Florence 1769

Bambach, Carmen C., *Drawing and Painting in the Italian Renaissance Workshop. Theory and Practice, 1300–1600*, Cambridge University Press, Cambridge 1999

Bambach, Carmen C. (ed. by), *Leonardo da Vinci Master Draftsman*, The Metropolitan Museum of Modern Art, New York 2003

Barasch, Moshe, *Light and Color in the Italian Renaissance Theory of Art*, New York University Press, New York 1978

Barasch, Moshe, "Personal Style: The Emergence of an Idea", *Biuletyn Historii Sztuk*, LIX, issue 3–4, 1997, pp. 180–188

Barasch, Moshe, *Theories of Art*, New York University Press, New York and London 1985

Barocchi, Paola (ed. by), *Scritti d'Arte del Cinquecento*, 3 vols, Ricciardi, Milan and Naples 1971–1977

Barocchi, Paola, *Trattati d'Arte del Cinquecento tra Manierismo e Controriforma*, LaTerza, Bari 1960–1962

Barozzi da Vignola, Iacopo, *Le Due Regole della Prospettiva Pratica*, Francesco Zanetti, Rome 1583

Bassi, Martino, *Dispareri in Materia d'Architettura e Prospettiva Con Pareri di Eccellenti e Famosi Architetti*, Marchetti, Brescia 1572

Baxandall, Michael, *Patterns of Intentions*, Yale University Press, New Haven 1985

Bayer, Andrea, *Defining Naturalism in Lombard Painting*, Yale University Press, New Haven 2004

Beat, Wismer; Dupré, Sven and Von Kerssenbrock-Krosigk, Dedo (eds.), *Kunst und Alchemie. Das Geheimnis der Verwandlung/Art and Alchemy, The Mystery of Transformation*, Hirmer Verlag, Munich 2014

Becker, Mira, "Grottesco & Suavitas: Zur Kopplung von ästhetischem Programm und institutioneller Form in zwei Mailänder Kunstakademien der Frühen Neuzeit", in Cancik-Kirschbaum, Eva and Traninger, Anita (eds.), *Wissen in Bewegung. Institution – Iteration – Transfer*, Wiesbaden: Harrassowitz Verlag 2016, pp. 415–440

Bellé, Riccardo, "Alcuni Aspetti dell'Ottica di Giovanbattista della Porta: dalla 'Magia Naturalis' al 'De Refractione'", extract from the conference *Lumen, Imago Pictura,* at the Bibliotheca Hertziana in Rome, 13th April 2010

Belloni, Luigi, *Storia della Medicina a Milano*, Treccani, Milan 1958–1962

Benedetti, Giovanni Battista, *Diversarum Speculationum Mathematicarum et Physicarum Liber*, apud Heredem Nicolai Bevilacqua, Turin 1585

Berra, Giacomo, "Alcune Puntualizzazioni sulla Pittrice Fede Galizia attraverso le Testimonianze del Letterato Gherardo Borgogni", *Paragone/Arte*, XL, 1989, pp. 14–29

Berra, Giacomo, "La Storia dei Canoni Proporzionali del Corpo Umano e gli Sviluppi in Area Lombarda alla Fine del Cinquecento", *Raccolta Vinciana*, XXV, 1993, p. 159–310

Berra, Giacomo, "Immagini Casuali, Figure Nascoste e Natura Antropomorfa nell'Immaginario Artistico Rinascimentale", *Mitteilungen des Kunsthistorischen Institutes in Florenz*, XLIII, 2–3, 1999, pp. 358–419

Berra, Giacomo, "Il Ritratto 'Caricato in Forma Strana, e Ridicolosa, e con Tanta Felicità di Somiglianza.' La Nascita della Caricatura e i suoi Sviluppi in Italia fino al Settecento", *Mitteilungen des Kunsthistorischen Institutes in Florenz*, LIII, 1, 2009, pp. 73–144

Bialostocki, Jan, "Doctus Artifex and the Library of the Artist in XVIth and XVIIth Century", in *De Arte et Libris, Festschrift Erasmus 1934–1984*, Amsterdam 1984, pp. 11–22

Biggam, C.P.; Hough, C.A.; Kay, C.J. and Simmons, D.R. (eds.), *New Directions in Color Studies*, John Benjamins Publishing Company, Amsterdam and Philadelphia 2011

Biondo, Michelangelo, *Della Nobilissima Pittura*, Bartolomeo l'Imperadore, Venice 1549

Blunt, Anthony, *Artistic Theory in Italy 1450–1600*, The Claredon Press, Oxford 1940

Bolzoni, Lina, *La Stanza della Memoria*, Einaudi, Turin 1995

Bolzoni, Lina, "Il Letterato come Giocatore e la Serietà del Gioco nelle *Carte Parlanti* di Pietro Aretino", *Renaissance Letters and Learning*, Warburg Institute, London 2012, pp. 7–21

Bonino, Cesare, *La Vita e i Miracoli di San Carlo Borromeo*, Jaca Book, Milan 2010

Bora, Giulio (ed. by), *Disegni di Manieristi Lombardi*, Neri Pozza, Vicenza 1971

Bora, Giulio, "La Prospettiva della Figura Umana. Gli 'Scurti' nella Teoria e nella Pratica Pittorica Lombarda del Cinquecento", in *La Prospettiva Rinascimentale. Codificazioni e Trasgressioni* ed. by Marisa Dalai Emiliani, Centro Di, Florence 1980, pp. 295–331

Bora, Giulio, "Da Leonardo all'Accademia de la Val di Bregno: Giovanni Paolo Lomazzo, Aurelio Luini e i Disegni degli Accademici", *Raccolta Vinciana*, XXIII, Milan 1989, pp. 73–101

Bora, Giulio, "Il Problema della Restituzione Prospettica: dal Metodo Geometrico agli Strumenti di Misurazione Empirica", *Arte Lombarda*, CX–CXI, 3–4, 1994, pp. 35–437

Bora, Giulio; Brown, David Alan and Carminati, Marco (eds.), *I Leonardeschi: l'Eredità di Leonardo in Lombardia*, Skira, Milan 1998

Bora, Giulio, Kahn-Rossi, Manuela and Porzio, Francesco (eds.), *Rabisch, Il Grottesco nell'Arte del Cinquecento*, Skira, Milan 1998

Bora, Giulio, *Toward a New Naturalism: Sixteenth-Century Painting in Cremona and Milan*, Yale University Press, New Haven 2004

Borghini, Raffaello, *Il Riposo*, Marescotti, Florence 1584

Borgogni, Gherardo, *La Fonte del Diporto*, Comin Ventura e Compagni, Bergamo 1598

Boschini, Marco, *Le Ricche Miniere della Pittura Veneziana*, Francesco Nicolini, Venice 1674

Bottari, Giovanni and Ticozzi, Stefano, *Raccolta di Lettere sulla Pittura, Scultura e Architettura*, 8 vols, A. Forni, Bologna 1979–1980

Britton, Piers, "*Mio Malinchonico, o Vero…mio Pazzo*: Michelangelo, Vasari, and the Problem of Artists' Melancholy in Sixteenth-Century Italy", *The Sixteenth-Century Journal*, XXXIV, 3, Autumn 2003, pp. 653–675

Buchmann, Jean, *Il Dialetto di Blenio*, E. Champion, Paris 1924

Bugati, Gasparo Domenico, *Historia Universale,* Eredi di Simone Tini, Milan 1579

Bury, Michael, *The Print in Italy 1550-1620*, The British Museum Press, London 2001

Calvesi, Maurizio, "Contribution de Gian Paolo Lomazzo à la Critique des Fiamminghi", *Les Arts Plastiques*, v, 1951, pp. 131–134

Canali, Paolo and Giorgi, Rosa, *San Carlo Borromeo e la Famiglia Francescana: Dialogo Fecondo tra Carisma e Istituzione*, Edizioni Biblioteca Francescana, Milan 2011

Carcano, Giovanni Battista, *De Vulneribus Capitis*, Pietro Tini, Milan 1584

Carcano, Giovanni Battista, *Exenteratio Cadaveris Illustrissimi Cardinalis Borrhomaei Mediolani Archiepiscopi, a Io. Baptista Carcano Leone, anatomen in Gymnasio ticiniensi publice profitente factae, verissima atque elegantissima enarratio*, Pietro Tini, Milan 1584

Cardano, Girolamo, *Opera Omnia*, 10 vols., Frommann, Stuttgart-Bad Cannstatt 1966

Cardi, Maria Virginia, "Intorno all'Autoritratto in Veste di Bacco di Giovan Paolo Lomazzo", *Storia dell'Arte*, LXXXI, 1994, pp. 182–193

Cardi, Maria Virginia, "Gian Paolo Lomazzo: La Cena Quadragesimale", *Bollettino Storico Piacentino*, I, 1996, pp. 77–89

Cardi, Lodovico, *Trattato Pratico di Prospettiva di Ludovico Cardi Detto il Cigoli: Manoscritto Ms 2660 A del Gabinetto dei Disegni e delle Stampe degli Uffizi*, Bonsignori, Rome 1992

Cazort, Mimi; Kornell, Monique and Roberts, K.B. (eds.), *The Ingenious Machine of Nature: Four Centuries of Art and Anatomy*, National Gallery of Canada, Ottawa 1996

Cennini, Cennino, *The Craftsman's Handbook: the Italian 'Il Libro dell'Arte'*, Dover Publications, New York 1933

Chai, Jean Julia, *Idea del Tempio della Pittura*, The Pennsylvania State University Press, University Park, Pennsylvania 2013

Chartier, Roger, *The Order of Books: Readers, Authors and Libraries in Europe Between the Fourteenth and Eighteenth Centuries*, translated by Lydia G. Cochrane, Stanford University Press, Stanford 1994

Cherubini, Francesco, *Dizionario Milanese-Italiano,* Imperial Regia Stamperia, Milan 1840

Ciardi, Roberto Paolo, "Struttura e Significato delle Opere Teoriche del Lomazzo", *Critica d'Arte*, XII, 70, 1965, pp. 20–30

Ciardi, Roberto Paolo, *Ambrogio Figino*, Marchi & Bertolli, Pisa 1968

Ciardi, Roberto Paolo, "A Knot of Words and Things: Some Clues for Interpreting the *Imprese* of Academies and Academicians", in *Italian Academies of the Sixteenth Century*, D.S. Chambers and F. Quiviger (eds.), London 1995, pp. 37–60

Ciardi, Roberto Paolo, "Struttura e Significato delle Opere Teoriche del Lomazzo", *Critica d'Arte*, XII–XIII, 1965, pp. 20–30 and pp. 37–74

Ciardi, Roberto Paolo, *Scritti sulle Arti*, Marchi e Bertolli, Pisa 1973

Clayton, Martin, *Leonardo da Vinci: the Divine and the Grotesque*, University of Chicago Press, Chicago 2002

Clark, Kenneth and Pedretti, Carlo (eds.), *Leonardo da Vinci's Anatomical Drawings at Windsor Castle*, Phaidon, London 1969

Clarke, Mark; De Munck, Bert and Dupré, Sven (eds.), *Transmission of Artists' Knowledge*, Koninklijke Vlaamse Academie van Belgie voor Wetenschappen en Kunsten, 2012

Clarke, Mark, *The Art of All Colors*, Archetype Publications, London 2001

Cojannot Le Blanc M.; Dalai Emiliani, M. and Dubourg-Glatigny, Pascal (eds.), *L'Artiste et l'Oeuvre à l'épreuve de la Perspective*, École Française de Rome, Rome 2006

Comanini, Gregorio, *Il Figino ovvero del Fine della Pittura*, Francesco Osanna, Mantua 1591

Consoli, Giuseppe, *I 'Giuochi' Borromeo ed il Pisanello*, Edizioni del Milione, Milan 1966

Cook, Harold and Dupré, Sven (eds.), *Translating Knowledge in the Early Modern Low Countries*, LIT Verlag, Zurich and Münster 2012

Coppa, Simonetta, "Federigo Borromeo Teorico d'Arte", *Arte Lombarda*, XV, 1970, pp. 65–70

Crossgrove, William, "The Vernacularization of Science, Medicine, and Technology in Late Medieval Europe: Broadening our Perspectives", *Early Science and Medicine*, V, 1, 2000, pp. 47–63

Cust, Lionel, "The Frescoes in the Casa Borromeo at Milan", *The Burlington Magazine*, XXXIII, 184, July 1918, pp. 8–14

D'Amico, Stefano, *Spanish Milan: a City Within the Empire, 1535–1706*, Palgrave MacMillan, New York 2012

Da Costa Kaufmann, Thomas, *Arcimboldo*, University of Chicago Press, Chicago 2010

Da Vinci, Leonardo, *Trattato della Pittura*, Giuseppe Brancato Editore, Catania 1990

Damm, Heiko; Thimann, Michael and Zittel, Claus (eds.), *The Artist as Reader. On Education and Non-Education of Early Modern Artists*, Brill, Boston and Leyden 2010

Daston, Lorraine, "Objectivity and the Escape from Perspective", *Social Studies of Science*, XXII, 4, November 1992, pp. 597–618

Daston, Lorraine, *The Nature of Nature in Early Modern Europe*, Max-Planck-Institut für Wissenschaftsgeschichte, Berlin 1997

Daston, Lorraine, *Wonders and the Order of Nature: 1150 - 1750*, Zone Books, New York 1998

Daston, Lorraine, *The Moral Authority of Nature*, University of Chicago Press, Chicago 2004

Daston, Lorraine, *Things That Talk: Object Lessons From Art and Science*, Zone Books, New York 2004

De Boer, Wietse and Göttler, Christine (eds.), *Religion and the Senses in Early Modern Europe*, Brill, Boston and Leyden 2013

De Munck, Bert, "Corpses, Live Models, and Nature: Assessing Skills and Knowledge before the Industrial Revolution", *Technology and Culture*, LI, 2, April 2010, pp. 332–356

Debolini, Francesco, "La Pittura Manierista di Aurelio Luini", Thesis of Specialization, University of Siena, 1995–1996

Delbourgo, James, "Listing People", *Isis*, IV, 103, 2012, pp. 735–42

Dempsey, Charles, "Some Observations on the Education of Artists in Florence and Bologna during the Later Sixteenth Century", *The Art Bullettin*, LXII, December 1980, pp. 552–569

Deswarte-Rosa, Sylvie, "Idea et le Temple de la Peinture: de Francesco de Holanda à Federico Zuccaro", *Revue de l'Art*, XCIV, 1991, pp. 45–65

Dolce, Lodovico, *Dialogo della Pittura di Lodovico Dolce Intitolato L'Aretino*, Giolito, Venice 1557

Dolce, Lodovico, *Dialogo dei Colori*, Gio. Battista Marchio, Venice 1565

Dubourg-Glatigny, Pascal and Vérin, Hélène (eds.), *Réduire en Art: La Technologie de la Renaissance aux Lumières*, Éditions de la Maison des Sciences de l'Homme, Paris 2008

Dupré Sven and Kusukawa, Sachiko (eds.), *The Circulation of News and Knowledge in Intersecting Networks*, Oxford University Press, Oxford 2008

Dupré, Sven, "The Dioptrics of Refractive Dials in the Sixteenth Century", *Nuncius. Annali di Storia della Scienza*, XVIII, 1, 2003, pp. 39–67

Dupré, Sven, "The Historiography of Perspective and Reflexy-Const in Netherlandish Art", *Netherlands Yearbook for History of Art/Nederlands Kunsthistorisch Jaarboek*, LXI, 2011, pp. 35–60

Dupré, Sven, "Ronchi's Leonardo", in *Leonardo '1952' e la Cultura dell' Europa nel Dopoguerra*, Romano Nanni and Maurizio Torrini (eds.), Biblioteca Leonardiana - Studi e Documenti, III, Olschki, Florence 2013, pp. 179–196

Dupré, Sven (ed. by), *Laboratories of Art: Alchemy and Art Technology from Antiquity to the 18th Century*, Archimedes, Springer, 2014

Dupré, Sven and Lüthy, Christoph (eds.), *Silent Messengers: The Circulation of Material Objects of Knowledge in the Early Modern Low Countries*, LIT Verlag, Berlin 2011

Dupré, Sven, "Kepler's Optics without Hypotheses", *Synthese*, CLXXXV, 3, 2012, pp. 501–525

Dupré, Sven; Kusukawa, Sachiko, and Maclean, Ian (eds.), *Transmitting Knowledge: Words, Images and Instruments in Early Modern Europe*, Oxford University Press, Oxford 2006

Dupré, Sven, *Renaissance Optics: Instruments, Practical Knowledge and the Appropriation of Theory*, Max-Planck-Institut für Wissenschaftsgeschichte Preprint 246, Berlin 2003

Eamon, William, *Science and the Secrets of Nature*, Princeton University Press, Princeton 1994

Eastlake, Charles, *Materials for a History of Oil Painting*, Longmans, London 1847

Equicola, Mario, *Dal Libro di Natura d'Amore*, Loranzo Lorio da Portes, Venice 1525

Eyferth, Jacob, "Craft Knowledge at the Interface of Written and Oral Cultures", *East Asian Science, Technology and Society*, IV, 2010, pp. 185–205

Fantoni, Marcello; Matthew, Louisa C. and Matthews-Grieco, Sara F. (eds.), *The Art Market in Italy 15th-17th Centuries*, Franco Cosimo Panini Editore, Ferrara 2003

Farra, Ferdinando Cesare, "Annotazioni Relative al Dialetto Usato dalla Cinquecentesca Accademia della Val del Blenio", *Rendiconti del (R.) Istituto Lombardo di Scienze e Lettere*, LXXXIV, 1951, pp. 289–307

Ferino-Pagden, Sylvia (ed. by), *Arcimboldo, Artista Milanese tra Leonardo e Caravaggio*, Exhibition Catalogue, Skira, Milan 2011

Field, J.V., "Giovanni Battista Benedetti on the Mathematics of Linear Perspective", *Journal of the Warburg and Courtauld Institutes*, XLVIII, 1985, pp. 71–99

Findlen, Paula, *Possessing Nature: Museums, Collecting and Scientific Culture in Early Modern Italy*, University of California Press, Berkeley 1994

Fiorani, Francesca and Nova, Alessandro (eds.), *Leonardo da Vinci and Optics*, Marsilio Editori, 2013

Frangenberg, Thomas, "Egnatio Danti's Optics. Cinquecento Aristotelianism and the Medieval Tradition", *Nuncius: Annali dell'Istituto e del Museo di Storia della Scienza di Firenze*, III, 1, pp. 3–38

Frangenberg, Thomas, "Il 'De Visu' di Giovan Battista Benedetti", in *Cultura, Scienze e Tecniche nella Venezia del Cinquecento*, Atti del Convegno Internazionle di Studio "Giovanni Battista Benedetti e il suo tempo" (Istituto Veneto di Scienze Lettere ed Arti, Venice), pp. 271–282

Frangenberg, Thomas, "Perspectivist Aristotelianism: Three Case-Studies of Cinquecento Visual Theory", *Journal of the Warburg and Courtauld Institutes*, LIV, 1991, pp. 137–158

Frangenberg, Thomas and Palmer, Rodney (eds.), *The Lives of Leonardo*, Warburg Institute Colloquia, London 2013

Freedberg, David, "The Failure of Colour", *Sight and Insight. Essays on Art and Culture in honour of E.H.Gombrich*, ed. by John Onians, Phaidon, London 1994, pp. 245–262

Gage, John, "Color in Western Art: An Issue?", *The Art Bulletin*, LXXII, 4, 1990, pp. 518–541

Gage, John, *Color and Meaning: Art, Science and Symbolism*, University of California Press, Berkeley and Los Angeles 1999

Galluzzi, Paolo, "Leonardo da Vinci: From the 'Elementi Macchinali' to the Man Machine", *History of Technology*, IV, 1997, pp. 235–265

Garfagnini, Gian Carlo and Smyth, Craig Hugh (eds.), *Florence and Milan: Comparisons and Relations*, Acts of Two Conferences at Villa I Tatti in 1982–1984, 2 vols, La Nuova Italia Editrice, Florence 1989

Gavel, Jonas, *Color: a Study of its Position in the Art Theory of the Quattro and Cinquecento*, Almqvist & Wiksell International, Stocholm 1979

Gengaro, Maria Luisa, "La Teoria dell'Arte di Giovan Paolo Lomazzo", *Archivio Storico Lombardo*, LIX, 1932, pp. 541–550

Gentilcore, David, *Medical Charlatanism in Early Modern Italy*, Oxford University Press, Oxford 2006

Giachino, Luisella, "La Lode e la Morte: Giuliano Goselini Poeta Funebre della Milano del Secondo Cinquecento", *Allegoria*, 2009/2010, XXVI, pp. 102–130

Giuliani, Marzia, "La Biblioteca di Casa Pellegrini", *Studia Borromaica*, XII, 1998, pp. 39–100

Godoy, José and Leydi, Silvio, *Parate Trionfali. Il Manierismo nell'Arte dell'Armatura Italiana*, 5 Continents Editions, 2003

Goldstein, Carl, *Teaching Art. Academies and Schools from Vasari to Albers*, Cambride University Press, Cambridge 1996

Goodrich, J.T., "Andreas Vesalius and Anatomy: a Re-evaluation of his Efforts", *Histoire des Sciences Médicales*, xvii, 1982, pp. 13–16

Goselini, Giuliano, *Rime di Diversi Celebri Poeti dell'Età Nostra*, Comin Ventura e Compagni, Bergamo 1587

Gottschalk, H.B., "The *De Coloribus* and its Author", *Hermes*, 1964, pp. 59–85

Grafton, Anthony, "Cardano and the Tradition of Classical Astrology. The Rothschild Lecture, 1995", *Proceedings of the American Philosophical Society*, CXLII, 3, September 1998, pp. 323–354

Grafton, Anthony, *Cardano's Cosmos. The Worlds and Works of a Renaissance Astrologer*, Harvard University Press, Cambrisge (Mass.) 1999

Grafton, Anthony, "Alberti's Terminology in Context", *I Tatti Studies*, VIII, 1999, pp. 37–68

Grendler, Paul, *Schooling in Renaissance Italy*, The Johns Hopkins University Press, Baltimore and London 1989

Hard, Frederick, "Some Interrelations Between the Literary and the Plastic Arts in 16th and 17th Century England", *College Art Journal*, X, 3, 1951, pp. 233–243

Harley, Rosamond D., *Artists' Pigments: A Study in English Documentary Sources (1600–1835)*, Butterworth Scientific, London 1970

Haydocke, Richard, *A Tracte Containing the Art of Curiose Painting* (Oxford 1598), Da Capo Press, Theatrum Orbis Terrarum LTD., Amsterdam and New York 1969

Heikamp, Detlef, "Vicende di Federigo Zuccari", *Rivista d'Arte*, XXXII, 1957, pp. 174–232

Hoeppe, Götz, *Why the Sky is Blue: Discovering the Color of Life*, Princeton University Press, Princeton 2007

Höltgen, Karl Josef, "Richard Haydocke: Translator, Engraver, Physician", *Library*, XXIII, 1978, pp. 15–32

Hope, Charles, "Sight Gags", *The New York Review of Books*, September 24, 1987, pp. 42–44

Hope, Charles, "Aspects of Criticsm in Art and Literature in Sixteenth-Century Italy", *Word and Image Conference Proceedings*, IV, 1–2, December-January 1988, pp. 1–10

Impey, Oliver and McGregor, Arthur, *The Origins of Museums: The Cabinet of Curiosities in Sixteenth- and Seventeenth- Century Europe*, House of Stratus, 2001

Ingegnoli, Vittorio, "Origine dell'Ambrosiana: Lelio Buzzi e Fabio Mangone", *Arte Lombarda*, X, 2, 1965, pp. 103–106

Isella, Dante, *Rabisch*, Einaudi, Turin 1993

Jones, Pamela, "Federico Borromeo's Ambrosiana Collection as a Teaching Facility for the Academy of Design", *Leids Kunsthistorisch Jaarboek*, V–VI, 1986, pp. 44–60

Jones, Pamela, *Federico Borromeo and the Ambrosiana. Art Patronage and Reform in Seventeenth-Century Milan*, Cambridge University Press, Cambridge (MA) 1993

Kemp, Martin, "'Equal Excellences': Lomazzo and the Explanation of Individual Style in the Visual Arts", *Renaissance Studies*, I, 1, March 1987, pp. 1–26

Kemp, Martin, *The Science of Art: Optical Themes in Western Art from Brunelleschi to Seurat*, Yale University Press, New Haven 1990

Kemp, Martin, "Lodovico Cigoli on the Origins and *Ragione* of Painting", *Mitteilungen des Kunsthistorischen Institutes in Florenz*, XXXV, 1, 1991, pp. 133–152

King, Catherine, "Late Sixteenth-Century Careers' Advice: a New Allegory of Artists' Training – Albertina, inv. No. 2763", *Viennaer Jahrbuch für Kunstgeschichte*, XLI, 1, pp. 77–96

Kirby, Jo, "The Painter's Trade in the Seventeenth Century: Theory and Practice", *National Gallery Technical Bulletin*, XX, 1999, pp. 5–49

Klein, Robert, *Idea del Tempio della Pittura*, Istituto Nazionale di Studi sul Rinascimento, 2 vols, Florence 1974

Klein, Robert, "Giudizio e Gusto dans la Théorie de l'Art au Cinquecento", *Rinascimento*, XII, 1, 1961, pp. 105–116

Klein, Robert, "Le Septs Gouverneurs de l'Art selon Lomazzo", *Arte Lombarda* IV, 2, 1959, pp. 277–287

Klein, Robert, *La Forme et l'Intelligible. Écrits sur la Renaissance et l'Art Moderne*, Gallimard, Paris 1970

Korrick, Leslie, *Lomazzo's Trattato della Pittura and Galilei's Fronimo: Picturing Music and Sounding Images in 1584*, Ashgate, Alershot 2003

Kwakkelstein, Michael, "Teste di Vecchi in Buon Numero", *Raccolta Vinciana*, XXV, 1993, pp. 38–62

Lamo, Alessandro, *Discorso di Alessandro Lamo Intorno alla Scoltura et Pittura*, Draconi, Cremona 1584

Laorca, Raffaello, *Le Tre Valli Stregate: Documenti per la Storia delle Streghe nei Baliaggi Svizzeri di Riviera, Blenio e Leventina: 1641–1676*, A. Dadò, Locarno 1992

Larry, Keith and Syson, Luke (eds.), *Leonardo da Vinci: Painter at the Court of Milan*, Exh. Cat., Yale University Press, New Haven and London, 2011

Laurenza, Domenico, "Figino and the Lost Drawings of Leonardo's Comparative Anatomy", *The Burlington Magazine*, CXLVIII, 1236, March 2006, pp. 173–179

Laurenza, Domenico, *Artists and Anatomy in Renaissance Italy*, The Metropolitan Museum of Art, New York 2012

Laurenza, Domenico, *De Figura Umana*, Olschki, Florence 2000

Lee, Rensselaer W., "Ut Pictura Poesis. The Humanistic Theory of Painting", *The Art Bulletin*, XXII (1940), pp. 197–269

Lencker, Hans, *Perspectiva Literaria: Das ist Ein Clerlicher Furreyssung* (1567), Kessinger Publishing, Whitefish, 2009

Leonhard, Karin, "Painted Poison: Venomous Beasts, Herbs, Gems and Baroque Colour Theory", *Art and Science in the Early Modern Netherlands*, Netherlands Yearbook for History of Art, lxi, 2011, pp. 116–147

Leonhard, Karin, *Bildfelder: Stilleben und Naturstücke des 17. Jahrhunderts*, Oldenbourg Akademieverlag, 2013

Leong, Elaine, "Collecting Knowledge for the Family: Recipes, Gender and Practical Knowledge in the Early Modern English Household", *Centaurus*, LV, 2, May 2013, pp. 81–103

Leong, Elaine and Rankin, Alisha (eds.), *Secrets and Knowledge in Medicine and Science, 1500–1800*, Ashgate, Farnham (Surrey) 2011

Lynch, James B., "Lomazzo's Allegory of Painting", *Gazette des Beaux Arts*, CX, 1968, pp. 325–330

Lyncoln, Evelyn, "The Studio Inventory of Camillo Graffio, Engraver and Fountainer", *Print Quarterly*, XXIX, 3, 2012, pp. 260–280

Lomazzo, Giovanni Paolo, *Gli Sogni e Raggionamenti, composti da Giovanni Paulo Lomazzo, Milanese, con le figure de' spiriti che gli raccontano, da egli dessignate*, Add MS 12196, Milan 1563

Lomazzo, Giovanni Paolo, *Trattato dell'Arte della Pittura, Scoltura et Architettura*, Paolo Gottardo Pontio, Milan 1584

Lomazzo, Giovanni Paolo, *Rime di Gio. Paolo Lomazzi, milanese pittore, diuise in sette libri: nelle quali ad imitatione de i grotteschi fatti da' pittori, ha cantato le lodi di Dio, et de le cose sacre, di prencipi, di signori, et huomini letterati, di pittori, scoltori, et architetti, et poi studiosamente senza alcun certo ordine, e legge accoppiato insieme vari et diuersi concetti tolti da filosofi, historici, poeti, et da altri scrittori*, Paolo Gottardo Pontio, Milan 1587

Lomazzo, Giovanni Paolo, *Rabisch*, Paolo Gottardo Pontio, Milan 1589

Lomazzo, Giovanni Paolo, *Idea del Tempio della Pittura*, Paolo Gottardo Pontio, Milan 1590

Lomazzo, Giovanni Paolo, *Della Forma delle Muse*, Paolo Gottardo Pontio, Milan 1591

Long, Pamela O., "Power, Patronage, and the Authorship of Ars: From Mechanical Know-How to Mechanical Knowledge in the Last Scribal Age", *Isis*, LXXXVIII, 1997, pp. 1–41

Long, Pamela O., *Artisan/Practitioners and the New Sciences 1400–1600*, Oregon State University Press, Corvallis 2011

Long, Pamela O., *Openness, Secrecy, Authorship: Technical Arts and the Culture of Knowledge from Antiquity to the Renaissance*, The Johns Hopkins University Press, Baltimore 2001

Longhi, Roberto, "Letteratura Artistica e Letteratura Nazionale", *Paragone*, XXXIII (1952), 33, pp. 7–14

Lynch, James, "Lomazzo and the Accademia della Valle di Bregno", *The Art Bulletin*, XLVIII, 1966, pp. 210–211

Mahon, Denis, *Studies in Seicento Art and Theory*, The Warburg Institute, London 1947

Manegold, Cornelia, *Wahrnehmung, Bild, Gedächtnis: Studien zur Rezeption der aristotelischen Gedächtnistheorie in den kunsttheoretischen Schriften des Giovanni Paolo Lomazzo*, Olms, Hildesheim 2004

Marinelli, Sergio, "The Author of the Codex Huygens", *Journal of the Warburg and Courtauld Institutes*, XLIV, 1981, pp. 214–222

Maspoli Genetelli, Silvia, *Il Filosofo e le Grottesche. La Pluralità dell'Esperienza Estetica in Montaigne, Lomazzo e Bruno*, Antenore, Roma-Padova 2006

Maylender, Michele, *Storia delle Accademie d'Italia*, 5 vols, Cappelli, Bologna 1998

Mayor, A. Hyatt, "Artists as Anatomists", *The Metropolitan Museum of Art Bulletin*, XXII, 6, February 1964, pp. 201–210

McGrath, Elizabeth, "Artists and Mythographic Handbooks: Some Evidence of Use and Ownership", *Images of the Pagan Gods*, Warburg Institute Colloquia, XIV, London 2009, pp. 389–419

McGrath, Elizabeth, "Artists, their Books, and Subjects from Mythology", in *Classical Mythology in the Netherlands in the Age of Renaissance and Baroque*, ed. by C. Van de Velde, Paris 2009, pp. 301–332

Meijer, Bert W., "From Leonardo to Brueghel: Comic Art in Sixteenth-Century Europe", *Word and Image*, IV, 1, 1988, pp. 405–411

Meijer, Bert W., *L'Arte non Deve Schernire: sul Comico e sul Grottesco al Nord*, Skira, Milan 1998

Melzi d'Eril, Giulio, "Federico Borromeo e Cesare Monti Collezionisti Milanesi", *Storia dell'Arte*, XV–XVI, 1972, pp. 293–306

Merrifield, Mary, *Original Treatises on the Art of Painting*, 2 vols, Dover Publications, New York 1967

Miedema, Hessel, "Karel van Mander: Did he Write Art Literature?", *Simiolus*, I–II, 22, 1993–1994, pp. 58–64

Migliaccio, Luciano, "Leonardo Auctor del Genere Comico", *Achademia Leonardi Vinci*, VIII, 1995, pp. 158–161

Milanesi, Gaetano, *Il 'Libro dell'Arte' o Trattato della Pittura*, F. Le Monnier, Florence 1859

Morandotti, Alessandro, "Pirro I Visconti Borromeo di Brebbia Mecenate nella Milano del Tardo Cinquecento", *Archivio Storico Lombardo*, CVII, 1984, pp. 115–162

Morandotti, Alessandro, "Nuove Tracce per il Tardo Rinascimento Italiano: il Ninfeo-Museo della Villa Borromeo, Visconti Boromeo, Litta, Toselli di Lainate", *Annali della Scuola Normale Superiore di Pisa*, Classe di Lettere e Filosofia, XV, 2, 1985, pp. 129–185

Morandotti, Alessandro, *Milano Profana Nell'Età dei Borromeo*, Milan, Electa 2005

Morato, Fulvio Pellegrino, *Del Significato dei Colori* (Mantua 1535), translated by Roy Osborne, Brown Walker Press, Florida 2012

Morigia, Paolo, *Della Nobiltà di Milano*, Pacifico Pontio, Milan 1595

Morigia, Paolo, *Historia dell'Antichità di Milano*, Appresso i Guerra, Venice 1592

Morigia, Paolo, (ed. by Borsieri, Girolamo), *Supplimento alla Nobiltà di Milano*, Pacifico Pontio, Milan 1619

Navoni, Marco, "L'Ambrosiana e il Museo Settala", in *Storia dell'Ambrosiana*, Cariplo, Milan 2000.

Neilson, Christina, *Between and Across Media: Verrocchio's Creation of Meaning in his Workshop*, PhD Dissertation, Johns Hopkins University, Baltimore 2009

Newman, William Royal and Principe, Lawrence M., "Alchemy vs Chemistry: The Etymological Origins of a Historiographic Mistake", *Early Science and Medicine*, III, 1, 1998, pp. 32–65

Nicodemi, Giorgio, "L'Accademia di Pittura, Scoltura e Architettura Fondata da Carlo Federigo Borromeo all'Ambrosiana", *Studi in Onore di Carlo Castiglioni, Prefetto dell'Ambrosiana*, Giuffré, Milan 1957, pp. 651–696

Noehles, Karl, *La Chiesa dei SS. Luca e Martina nell'Opera di Pietro da Cortona*, Ugo Bozzi Editore, Rome 1970

Nova, Alessandro, "Pietre Naturali, Gessi Colorati, Pastelli e il Problema del Ritratto", *Mitteilungen des Kunsthistorischen Institutes in Florenz*, LII, 2–3, 2008, pp. 158–175

Occolti, Coronato, *Trattato de' Colori*, Appresso Seth Viotto, Parma 1568

Olmi, Giuseppe, *Rappresentare il Corpo: Arte e Anatomia da Leonardo all'Illuminismo*, Bononia University Press, Bologna 2004

Olszewski, Edward J., "Distortions, Shadows and Conventions in Sixteenth-Century Italian Art", *Artibus et Historiae*, VI, 11, 1985, pp. 101–124

Omodeo, Pietro Daniel, "La Cosmologia Infinitistica di Giovanni Battista Benedetti", *Bruniana & Campanelliana*, XV, 1, 2009

Orlandi, Pellegrino Antonio, *Abcdario Pittorico*, Constantino Pisarri, Bologna 1719

Ossola, Carlo, *Autunno del Rinascimento: Idea del Tempio dell'Arte nell'Ultimo Cinquecento*, L.S. Olschki, Florence 1971

Paliaga, Franco, "Giovanni Ambrogio Brambilla, le 'Teste di Carattere' di Leonardo e la Commedia dell'Arte", *Raccolta Vinciana*, XXVI, 1995, pp. 219–254

Panofsky, Erwin, *The Codex Huygens and Leonardo da Vinci's Art Theory*, Studies of the Warburg Institute, xiii, ed. by Fritz Saxl, London 1940

Panofsky, Erwin, *Idea: ein Beitrag zur Begriffsgeschichte der älteren Kunsttheorie*, Studien der Bibliothek Warburg, v, Leipzig-Berlin 1924

Park, Katharine, "The Criminal and Saintly Body", *Renaissance Quarterly*, XLVII, 1, Spring 1994, pp. 1–33

Parma, Elena, *Perin del Vaga: L'Anello Mancante*, Sagep, Genoa 1986

Pavesi, Mauro, *Giovanni Paolo Lomazzo Pittore Milanese 1538–1592*, PhD Dissertation at the Università Cattolica del Sacro Cuore, Milan 2006–2007

Pavesi, Mauro, "Qualche Riflessione sull'Attività Pittorica di Giovanni Paolo Lomazzo", in *Studi in Onore di Maria Grazia Albertini Ottolenghi*, M. Rossi, A. Rovetta, F. Tedeschi (eds.), Milan 2013, pp. 155–161

Payne, Alina, "Vasari, Architecture and the Origins of Historicizing Art", *RES*, XL, 2001, pp. 51–76

Pedretti, Carlo, "Leonardo's Allegories at Oxford Explained by Lomazzo", *The Burlington Magazine*, XCVI, 615, 1954, pp. 175–177

Pedretti, Carlo, *The Literary Works of Leonardo da Vinci: Compiled and Edited from the Original Manuscripts by Jean Paul Richter*, University of California Press, Berkeley and Los Angeles, 1977

Pedretti, Carlo. "Leonardo a Venezia", *Giorgione 1478–1978: Guida alla Mostra. I Tempi di Giorgione Series*, ed. by P. Carpeggiani, Alinari, Florence 1978

Pedretti, Carlo, *Leonardo: Studies for the Last Supper from the Royal Library at Windsor Castle*, Electa, Milan 1983

Pedretti, Carlo, "L' Altro Leonardo", in *Fra Rinascimento, Manierismo e Realtà*, Scritti di Storia dell' Arte in Memoria di Anna Maria Brizio, ed. by Pietro C. Marani, Giunti Barbera, Florence 1984

Pellegrini, Pellegrino, *L'Architettura*, ed. by G. Panizzi and A. Buratti Mazzotta, Polifilo, Milan 1990

Pevsner, Nikolaus, *Academies of Art Past and Present*, Cambridge University Press, Cambridge 1940

Pfisterer, Ulrich, *Kunst-Geburten: Kreativität, Erotik, Körper*, Verlaug Klaus Wagenbach, Berlin 2014

Picinelli, Filippo, *Ateneo dei Letterati Milanesi*, Francesco Vigone, Milan 1670

Pino, Paolo, *Dialogo della Pittura*, Paolo Gherardo, Venice 1548

Popham, Arthur Ewart, *Italian Drawings Exhibited at the Royal Academy*, Oxford University Press, London 1931

Popham, Arthur Ewart, "On a Book of Drawings by Ambrogio Figino", *Bibliothèque d'Humanisme et Renaissance, Travaux et Documents*, XX, 1958, pp. 266–276

Priorato, Galeazzo Gualdo, *Relazione della Città e dello Stato di Milano*, appresso Lodovico Monza, Milan 1666

Ramelli, Agostino, *Le Diverse et Artificiose Macchine,* in Casa dell'Autore, Paris 1588

Raponi, Nicola and Turchini, Angelo eds., *Stampa, Libri e Letture a Milano nell'Età di San Carlo Borromeo,* Vita e Pensiero, III, Milan 1982

Redi, Ladislao, *The Library of Leonardo da Vinci,* The Castle Press, Pasadena, California 1972

Richards, Jennifer, "Useful Books: Reading Vernacular Regimens in Sixteenth-Century England", *Journal of the History of Ideas,* LXXIII, 2, 2012, pp. 247–71

Richter, Irma, "The Codex Huygens", review of Panofsky's book in *The Art Bulletin,* XXIII (1941), pp. 335–338

Rijks, Marlise, 'Defenders of the Image. Painted Collectors' Cabinets and the Display of Display in Counter-Reformation Antwerp', in *Arts of Display. Netherlands Yearbook for History of Art,* H. Perry Chapman, F. Scholten, and J. Woodall (eds.), pp. 54–83

Roccasecca, Pietro, "Vincenzo Foppa: Evidenze Materiali della Procedura Prospettica nei Tre Crocifissi dell'Accademia Carrara di Bergamo", in *Vincenzo Foppa: Tecniche d'Esecuzione, Indagini e Restauri,* Skira, Milan 2002, pp. 59–67

Roccasecca, Pietro, "La Prospettiva Lineare nel Quattrocento dalle Proporzioni Continuata e Ordinata alla Proporzione Degradata", in *Proportions. Science, Musique, Peinture & Architecture,* Brepols Publisher, Turnout 2011, p. 277–297

Roccasecca, Pietro, "Il Ruolo della Teoria Düreriana nella formazione degli artisti nell'Accademia del Disegno di Roma", in *Dürer, l'Italia e l'Europa,* Sybille Ebert-Schifferer, Kristina Herrmann Fiore (eds.), Cinisello Balsamo, Milan 2011, pp. 88–97

Rocchi, Giuseppe, "La Diffusione dei Modi Rinascimentali a Milano: Esemplicazioni", *Arte Lombarda,* LXXXVI–LXXXVII, 1988, pp. 83–93

Romano, Giovanni, *Rinascimento in Lombardia: Foppa, Zenale, Leonardo e Bramantino,* Feltrinelli, Milan 2011

Rosenberg, Charles M. (ed. by), *The Court Cities in Northern Italy: Milan, Parma, Piacenza, Mantua, Ferrara, Bologna, Urbino, Pesaro and Rimini,* Cambridge University Press, Cambridge 2010

Rossi di Marignano, Federico, *Carlo Borromeo, un Uomo, una Vita, un Secolo,* Mondadori, Milan 2010

Ruffino, Alessandra (ed. by), *Rime ad Imitazione de i Grotteschi Usati da' Pittori,* Vecchiarelli Editore, Manziana (Rome) 2006

Saba, Agostino, "La Biblioteca di San Carlo", *Fontes Ambrosiani,* XII, Florence 1936, pp. 13–20

Sandal, Ennio, *Editori e Tipografi a Milano nel Cinquecento,* 3 vols, Korner, Baden-Baden 1977–1978

Scamozzi, Vincenzo, *Idea dell'Architettura Universale,* Girolamo Abrizzi, Venice 1714

Scaramuccia, Luigi, *Le Finezze dei Pennelli Italiani,* Magni, Pavia 1674

Schlosser, Julius Ritter Von, *La Letteratura Artistica: Contributi alla Bibliografia delle Fonti della Storia dell'Arte,* La Nuova Italia, Florence 1937

Schmandt-Besserat, Denise (ed. by), *Early Technologies*, Undena Publications, Malibu 1979

Schmidt, Benjamin and Smith, Pamela, *Making Knowledge in Early Modern Europe: Practices, Objects, and Texts, 1400–1800*, University of Chicago Press, Chicago 2008

Schmitt, Lothar, "Education and Learning among Sixteenth-Century German Artists", *Studies in the History of Art*, LX, Symposium Papers XXXVII, "Hans Holbein: Paintings, Prints, and Reception", 2001, pp. 72–81

Scholl, Dorothea, *Von den 'Grottesken' zum Grotesken: die Konstituierung einer Poetik des Grotesken in der Italienischen Renaissance*, Ars Rhetorica 11, Muenster 2004

Schultz, Bernard, *Art and Anatomy in Renaissance Italy*, UMI Research Press, Ann Arbor (Mich.) 1985

Sénécal, Robert, "Carlo Borromeo's Instructiones et Suppellectilis Ecclesiasticae and its Origins in the Rome of his Time", *Papers of the British School at Rome*, LXVIII, 2000, pp. 241–267

Shapiro, Alan E., "Artists' Colors and Newton's Colors", *Isis*, LXXXV, December 1994, pp. 600–630

Shell, Janice, *Pittori in Bottega*, Allemandi, Turin 1995

Sinisgalli, Rocco (ed. by), *Il Nuovo de Pictura di Leon Battista Alberti*, Edizioni Kappa, Rome 2006

Siraisi, Nancy G., *Medicine and the Italian Universities*, Brill, Leiden 2001

Smith, Pamela, "Sixteenth-Century Audiences for How-To: Who Read How-To Books?", conference paper, *Reading How-To Workshop*, Max Planck Institute for the History of Science, Berlin, 19–20 September 2014

Smith, Pamela, "The History of Science as a Cultural History of the Material World", in *Cultural Histories of the Material World*, ed. by Peter Miller, University of Michigan Press, Ann Arbor 2013, pp. 210–225

Solmi, Edmondo, "Ricordi della Vita e delle Opere di Leonardo", *Archivio Storico Lombardo*, XXXIV, 1907, pp. 290–331

Sorte, Cristoforo, *Osservazioni Nella Pittura*, Girolamo Zenaro, Venice 1580

Speakman Sutch and Van Dixhoorn (eds.), *The Reach of the Republic of Letters*, 2 vols, Brill 2008

Spina-Barelli, Emma, "Il Lomazzo o il Ruolo delle Personalità Psicologiche nell'Estetica dell'Ultimo Rinascimento", *Arte Lombarda*, III, 2, 1958, pp. 119–124

Stefani, Stefania, "Idea dello Stile della Pittura di Giov. Paolo Lomazzo", *Acme*, XVI, 2, 1963, pp. 41–56

Swan, Claudia, "Collecting Naturalia in the Shadow of Early Modern Dutch Trade", in *Colonial Botany. Science, Commerce, and Politics in the Early Modern World*, ed. by Londa Schiebinger and Claudia Swan, University of Pennsylvania Press, Philadelphia 2004, pp. 223–236

Swan, Claudia, "Memory's Garden and Other Wondrous Excerpts: Ernst Brinck (1582–1649), Collector", *Kritische Berichte*, III, 2012, pp. 5–19

Taddeo, Edoardo, "I Grilli Poetici di un Pittore: le Rime di G.P. Lomazzo", *Il Contesto*, III, 1997, pp. 140–181

Tavernari, Carlo, "Manfredo Settala, Collezionista e Scienziato Milanese del '600", *Istituto e Museo di Storia della Scienza di Firenze*, I, 1, 1976, pp. 43–61.

Tea, Eva, "Un Pittore Tomista: Giovanni Paolo Lomazzo", *Rivista Filosofica Neoscolastica*, XXXI, 4 and 5, 1939, pp. 408–419

Terpening, Ronnie H., *Lodovico Dolce: Renaissance Man of Letters*, University of Toronto Press, Toronto 1997

Tettamanzi, Dionigi, *San Carlo e la Croce*, Ancora, Milan 2010

Theophrastus (Pseudo-Aristotle), *De Coloribus*, 793b34–794a15; in Aristotle, *Minor Works*, transl. by W.S. Hett, Loeb Classical Library, Harvard University Press, Cambridge (Mass.) 1952

Tiraboschi, Girolamo, *Storia della Letteratura Italiana*, 9 vols, Molini, Landi and Co., Florence 1805–1813

Tomai, Tomaso, *Idea del Giardino del Mondo*, Pacifico Da Ponte ad instanza haer. Pietro Tini, Milan 1589

Tramelli, Barbara, "Due Poesie del Pittore Giovanni Paolo Lomazzo, Ammiratore di Cardano", *Bruniana et Campanelliana*, XVI, II, 2010, pp. 573–576

Tramelli, Barbara, "Artists and Knowledge in Sixteenth-Century Milan: The Case of Lomazzo's Accademia de la Val di Blenio", *Art and Knowledge -Special issue of Fragmenta*, Journal of the Royal Netherlands Institute in Rome, 2011, pp. 121–137

Trutty-Coohill, Patricia, "Making the Dead Laugh", *Achademia Leonardi Vinci*, X, 1997, pp. 190–196

Valerio, Anna Patrizia, "Annibale Fontana e il Paliotto dell'Altare della Vergine dei Miracoli in Santa Maria presso San Celso", *Paragone*, XXIV, 279, 1973, pp. 32–53

Van Ringelbergh, Joachim Sterck, *Liber de Ratione Studii*, Bartholomeus Westheimer, Basel 1541

Van Son, Rogier, "Lomazzo, Lampsonius en de Noordelijke Kunst", *Nederlands Kunsthistorisch Jaarboek*, XLIV, 1, 1993, pp. 185–196

Varchi, Benedetto, *Lezzione della Maggioranza delle Arti*, Giunti, Florence 1590

Vasari, Giorgio, *Vite de' più Eccellenti Pittori, Scultori e Architetti Moderni*, Florence 1568, ed. by R. Bettarini and P. Barocchi, Sansoni, Florence 1967

Vérin, Hélène, *Le Gloire des Ingénieurs: L'Intelligence Technique du XVIe au XVIIIe Siècle*, Albin Michel, Paris 1993

Vredeman de Vries, Hans, *Artis Perspectivae Plurium Generum Elegantissimae Formulae antea numquam Impressae*, Gerardus de Iode, Neomagensis, Antwerp 1568

Vredeman de Vries, Hans, *Perspective, Das ist Die weitberuembte khunst, eines scheinenden in oder durchsehenden augengesichts Puncten*, Hondius, Leiden 1604

Wallert, Arie, "Academy and Workshop; Mediaeval Transmission of Antique Knowledge', in *Making and Transforming Art: Changes in Artists", Materials and Practice*,

Proceedings of the fifth symposium of the Art Technological Source Research Working Group, H. Dubois, S. Eyb-Green, J.H. Townsend (eds.), London 2013

Warwick, Genevieve, "Collecting as Canon Formation. Art History and the Collection of Drawings in Early Modern Italy", in *Memory and Oblivion*, Reinik Wessel and Jeroen Stumpel (eds.), Dordrecht 1999, pp. 191–204.

Welch, Evelyn, *Art and Authority in Renaissance Milan*, Yale University Press, New Haven 1995

Welch, Evelyn, *Shopping in the Renaissance*, Yale University Press, New Haven and London 2005

Werrett, Simon, *Fireworks: Pyrotechnic Arts and Science in European History*, University of Chicago Press, Chicago 2009

West Fitzhugh, Elisabeth (ed. by), *Artists' Pigments: A Handbook of Their History and Characteristics*, 3 vols, National Gallery of Art, Washington 1986–1997

Weststeijn, Thijs, *The Visible World*, Amsterdam University Press, Amsterdam 2008

Williams, Robert, *Art, Theory, and Culture in Sixteenth-Century Italy: From Techne to Metatechne*, Cambridge University Press, Cambridge 1997

Wittkower, Rudolph and Margot, *Born under Saturn*, Weidenfeld & Nicolson, London 1963

Wood, Christopher S., "The Perspective Treatise in Ruins: Lorenz Stoer, *Geometria et Perspectiva*, 1567", *Studies in the History of Art*, XLIX, Symposium Papers XXXVI, The Treatise on Perspective: Published and Unpublished, 2003, pp. 234–257

Young Kim, David, *The Traveling Artist in the Italian Renaissance: Geography, Mobility and Style*, Yale University Press, New Haven 2014

Zardin, Danilo, *Carlo Borromeo, Cultura, Santità e Governo*, Vita e Pensiero, Milan 2010

Zuccari, Federico *Il Passaggio per l'Italia con la Dimora di Parma del Cavalier Federico Zuccaro* (1603), ed. by V. Lanciarini, Rome 1893

Zuccari, Federico, *Origine e Progresso dell'Accademia del Disegno a Roma*, Pietro Bartoli, Pavia 1604

Index of Names

www.ingramcontent.com/pod-product-compliance
Lightning Source LLC
LaVergne TN
LVHW010626300926
849162LV00001B/73